Praise for *Fifteen Million Years in Antarctica*

'It's a book about learning, understanding and communication, but also about rapture and dread and awe. It's an important book, and one that is a joy to read.' —Elizabeth Knox

'An utterly engrossing, surprisingly relatable memoir combining science, awe, anxiety, family life – and the spectre of climate-change devastation.' —*NZ Listener*, Best 100 Books of 2019

'Priestley's intelligence is a curious mix of creative and scientific. She cleverly brings the two worlds together, humanising scientists and making their stories and discoveries accessible.' —Sarah Catherall, *Sunday*

'This is a book of multiple discoveries. Self discovery, geological discovery, planet discovery – and the more you read the more you determine choices that need to be made.' —Paula Green

'Antarctica has a pivotal moderating role in global climate. Few science writers get there and fewer still connect as personally and as well as Priestley.' —Neville Peat, *Landfall Review Online*

'a deft eco-memoir . . . This is a book about personal as well as global anxiety, about geology, about memory and connection, about the science of climate change. Most pressingly, it is a book that reminds us of all that we stand to lose if we don't change our priorities.' —*NZ Listener*

'Generous, authoritative and accessible.' —David Hill, *Weekend Herald*

'Refreshing, direct and unhindered by political niceties. Her language is open, expressive. . . . This book does not hold back its anger or its tears.' —Bob Frame, *Polar Record*

'Priestley is unafraid to confront the truths of our desecration of the planet, led by irresponsible and ignorant politicians. It is (pun intended) a chilling read, but also a very important one. Priestley's triumph is to engage us with this inhospitable wasteland, and to help us see what we are doing to it. Essential reading!' —Steve Walker, *Dominion Post*

End Times

Rebecca Priestley

TE HERENGA WAKA
UNIVERSITY PRESS

Te Herenga Waka University Press
PO Box 600 Wellington
New Zealand
teherengawakapress.co.nz

ISBN 9781776921188

A catalogue record is available at the National Library
of New Zealand.

Printed in Singapore by Markono Print Media Pte Ltd

For Maz, my BFF

Contents

Everyone's talking about World War Three
But we're as safe as safe can be
There's no unrest in this country

Blam Blam Blam, 'There is no depression in New Zealand'

1

Faultline

'Where's the Alpine Fault from here?' asks Maz, a finger hovering over the map.

Glacier Country – untamed natural wilderness it says at the top. There's a key for restaurants, shops, hotels and walking tracks. Franz Josef has two main streets, one being the state highway we drove in on, with a few side streets connecting the two. For our last night on the road, we're staying at an 'eco retreat' – a scattering of cabins and tree-houses – and we've chosen a single-storey, bush-surrounded cabin.

The host points to a junction between the highway and one of the side streets. 'Just up here,' she says in a lovely Irish accent, 'through the Mobil station.'

Adrenaline floods my body. 'What?' I say. '*What?*'

Alpine Fault is a game my best friend Maz and I have been playing all the way down State Highway 6. How fast can we drive over the plate boundary to minimise the risk of being swallowed by an Alpine Fault rupture? What would happen if the Alpine Fault moved *right now*, while we're crossing this bridge? Whose house would we crash at if an Alpine Fault earthquake cut the West Coast off from the rest of the country? It's been fun, tapping into a deep-seated survivalist tendency we both have.

But now, without thinking it through, I've paid for a night in a cabin 300 metres from the Alpine Fault trace. What the fuck? How have I missed this? It's an unnecessary risk. An Alpine Fault movement while we're here wouldn't just mean being cut off from the rest of the country; it would mean an immediate risk to life. I feel like an idiot. I'm okay if I have a plan, but if we're going to be swallowed by a plate boundary rupture, I don't have an escape strategy.

Maz is looking at me and laughing. She says that everyone on the West Coast knows the Alpine Fault runs through Franz Josef. Surely I knew this? Aren't I the one with the geology degree? The host, smiling but looking perplexed, locks eyes with Maz and hands her the key.

I'm mute with anxiety. I rethink our plan. Should I insist we leave town or move to a hotel further away from the fault? I start thinking about statistics, and how likely it is that the Alpine Fault will rupture tonight. It's a low-probability event but extremely high stakes. While I'm thinking, we get back in the car and Maz drives us to our cabin.

The cabin is cute – a modern wood-and-corrugated-iron box with a sliding front door and a single-pitch roof over a front porch. We talk about what would happen in an Alpine Fault earthquake. From a quick literature review on my phone, I've ascertained that we're not in the town's 130-metre-wide fault rupture avoidance zone, a high hazard area in which there's a chance of bending, folding and rupture of up to two metres vertically and eight metres horizontally, but we're fucking close to it. Where we are, we could expect intense shaking for up to four minutes. We would likely be thrown to the floor, but our cabin – made of ductile materials – would move with the motion rather than collapse. So, as long as we duck and cover, and nothing falls on us, we should be okay.

Once we've checked out the cabin construction, I stand on the front porch and look up at the bush-clad slopes just a few hundred metres away. Shaking of the intensity you'd get from an Alpine Fault rupture causes widespread landslides, rockfalls, liquefaction. *The hill slopes to the east of Franz Josef are characterised by steep range front topography and weak fault-crushed rock mass that when combined with high rates of rainfall and seismicity equate to high landslide susceptibility,* says a report I find online.

'Are you wondering if those hills are going to come down?' says Maz.

'Yep.'

'Hmmm,' she says. 'I clocked that bulge as we were driving into town.'

On the slope above us is a distinctive bulge. Maz starts muttering about angle of repose, and we talk it all through. Eventually we decide that, given the Alpine Fault goes every 300 years or so, there's not a lot of regolith there to come down in the inevitable landslides that a massive quake would cause. So, as long as the hill doesn't *totally* come down on us in a massive rock avalanche, we should survive. I'm somewhat reassured, but I need to feel more prepared, so I head back into town to stock up on emergency supplies.

I drive to the Four Square – avoiding the bump in the road near the Mobil station – and get water, tuna, crackers, cheese, dried fruit and chocolate. Back at the cabin, I empty the books, knitting and kids' presents from my backpack and refill it with food, bottled water and a roll of loo paper. I put our sleeping bags and parkas near the door. There's no point having them in the car if it gets crushed by a tree or the road is blocked.

It starts raining. I have a folkloric sense of earthquake weather being hot, humid and still, so the rain feels comforting

– slightly more cosy and safe. But this is flawed reasoning, because wet ground increases the risk of earthquake-triggered landslides.

In the square cabin there's a double bed, a single bed, a coffee table with a couch and comfortable chairs, and a bathroom. I throw my things onto the single and let Maz, the restless sleeper, take the big bed. My go bag prepared, and committed to being here now, I try to settle into our digs.

But I'm still anxious. Maz checks out the drinks in the fridge. It's good news: there are local beers and wines and they're all complimentary. We've paid a lot for this place so we make a commitment to drink them all. We start with a West Coast Red IPA; we'll have the wine after dinner.

On the shelf above the bench is a ceramic jar full of incense sticks. Maz takes an electric fire-lighter from the bench and tries to light one. It won't take, so she goes to sit on the porch with her beer. We try to settle into a mellow mood, though I'm still on edge. While I'm drinking a beer, I roll a big fat spliff to help me relax. I haven't done this for decades but I have a body memory of what to do.

We can't smoke it yet, though, because Maz has a Zoom meeting and I need to scarper.

In the dusk, I drive out of town, far enough out to feel safe. I look at the proximity of the hills, the river, the floodplain, and check between the map and what I can see out the window. I drive over Tartare Stream, then panic when I realise the road is taking me closer to the mountains. I go over Stony Creek to a spot where the road curves north around a gentle hillock, away from the mountains. Flat land is ahead. I decide this is a 'safe' spot we could drive to if we were hit by a quake and the roads out of town were open. Part of the plan.

―

14

When I was at primary school, and spent most lunchtimes inside reading, one of my fears was spontaneous human combustion. Funny how the things you worry about change as you get older. But I realise that, in one way or another, I've been poised for disaster, catastrophe, apocalypse even, my entire life. The End Times have always seemed imminent, and I've been perpetually alert for tectonic ruptures, Biblical raptures, nuclear blasts, the metaphorical ticks of the Doomsday Clock's second hand. Through the decades, the End Times have travelled along with me, taking on new forms, gathering in new fears.

Lately, it feels as though the End Times have arrived. That adrenaline-fuelled night on the Alpine Fault I was kept awake by a fear of being swallowed by the earth, by my disbelief that no one else seemed to be taking the danger seriously. But most nights, when I'm not sleeping over a plate boundary, or checking my symptoms for signs of a deadly virus, the thing that keeps me awake is climate change.

2

Methane seep

Maz's mother says that when we first met, as a pair of two-year-olds, we looked directly into each other's eyes. It was a look of recognition, she says. Then we toddled into each other's arms. We've been friends ever since. Our fathers worked together, as engineers; our mothers were primary school teachers; our grandfathers taught at the same high school; our great- and great-great-grandfathers were gold miners. But we think there's something deeper. We've followed different paths and have sometimes gone years without seeing each other, but we always come back to each other.

At the start of our road trip, about a week before my sleepless night at Franz Josef, Maz and I met up in Nelson after flying in from Auckland and Wellington. Our decades-long friendship has endured different career paths, different home lives, many years spent in different cities and countries. Now, we were having a moment of synchronicity. Maz, a mining engineer turned civil engineer, was recently separated from her husband, and I was disillusioned with my job. Maz needed a holiday, and I needed a new focus, a writing project. I was eight years into an academic career after twenty years of self-employment and study, and I was under instruction from my coach, who had been supporting me through some work issues, to 'fucking enjoy yourself'. Maz and I thought a week-long roadie would meet both of our needs.

A few months earlier, the Bulletin of the Atomic Scientists had set the Doomsday Clock, which measures how close we are to destroying the world with dangerous technologies of our own making, sat at 100 seconds to midnight – the closest since 1953. Their announcement headlined the COVID-19 pandemic, pointing out how 'unprepared and unwilling' the international community was to deal with a global emergency, and naming nuclear weapons and climate change as posing 'existential threats to humanity'.

After a career spent interviewing scientists about these issues, along with other horrors, I was keen to hear from other folk. Including those who were literally at the coal face, working in mines. I knew there were people on the West Coast with beliefs and worldviews wildly different from those of my middle-class, liberal, Wellington community, and I wanted to talk to them – I wanted to *listen* to them – and hear what they thought about the scientific issues that are sometimes described as controversial but to me are no-brainers.

The rekindling of my friendship with Maz started early that year too, at a writing retreat during which Maz wrote about her time as a mining engineer in Australia and I scoured my teenage journals to help me into an experimental piece of writing. I'd wanted to 'be a writer' since I was a teenager, and my journals were detailed and voluminous. I trawled through entries from 1983 to 1985, halfway through sixth form to the end of my gap year. In the evenings we drank wine in my room and I read aloud and we laughed and connected over the dark intensity of those years, marvelling at our righteous and judgemental teenage selves.

Now we were on the road, equipped with sleeping bags, books, sturdy walking boots. I was uncertain which clothes to bring and decided to go for a bit of a Lumberjane look – jeans, checked shirts, boots. While I like black clothes, big sunglasses and elegant coats, I'm also inclined to scruffiness. Maz has a definite style. Her dark hair is bleached blond and cut into a short choppy bob, she has the smooth forehead and plump lips of a high-salaried Aucklander, and she favours expensively ripped designer jeans and skull T-shirts. When she told me she was packing 'Normcore' for this trip, I went to my local Farmers to buy a black merino hoodie and a pair of stretchy, dark green, slim-leg pants. But was any of this really me?

When Maz and I were teenagers, we were always getting dressed up. Our costumes included Auntie Maudie's Victorian lace underwear, Uncle Jack's oversized army jackets, 1940s floral dresses, New Romantics pirate layers, punk garb and even flouncy peach-coloured choir dresses. Sometimes we dressed like the singers we liked from Split Enz, Ultravox, Adam and the Ants, the Sex Pistols. Sometimes we would hold hands as we walked down the main street of Petone or Whakatāne to try and fuck with any homophobic locals. Back then, you didn't need to do much to turn heads.

During lockdown, I let the grey streaks grow through my hair, stopped wearing the little makeup I usually applied, spent most days in gardening clothes. I wondered if my academic career was a kind of performance too – one that I felt too tired to keep doing. But without it, without my costume, I felt untethered. When I looked in the mirror, I didn't know what to make of my reflection.

⁓

Day one. Late autumn. In Wakefield we park on the gravel verge, across the highway from the Wakefield pie shop, behind a heavily laden white ute. On the back tray is an enormous, black, three-headed beast, foul and Biblical, defying all laws of nature. I put on my glasses, take a closer look. Through a small window in a plywood dog box hang three black and tan heads, grinning mouths open, teeth showing, tongues dripping. Draped over the top of the box is an enormous wild pig, a hairy old Captain Cooker. Welcome to rural Aotearoa.

Inside the shop we inspect a double-sized warmer filled with hot pies – mince and cheese, steak and kidney, steak and mushroom, bacon and egg, and more. I ask about the vegetarian options – there are two – then choose the vegetable pie.

'Leek, celery, broccoli, cauliflower bound in a cheese sauce with a slice of kūmara and mashed potato on top. I really rate it,' says the woman behind the counter.

Maz – who knows the shop, used to come here regularly – chooses a steak, stout and mushroom pie.

'These are the best fucking pies,' she says, then takes a bite from the paper-bagged pastry.

From Wakefield we take our hybrid rental east along State Highway 6, through pine forests misty with layers of cloud and over swollen brown rivers. Maz drives fast while I sit in the passenger seat, operating the sound system and taking notes. She's driven this road dozens of times, from Westport to Nelson for lunches, shopping, visits with friends.

I've made a playlist, 41 songs that cover the formative years of our friendship between the ages of 13 – when we found ourselves in the same class at high school – and 18, when Maz went to Auckland to study engineering. It starts with the Topp Twins 'Untouchable Girls', with our favourite line: *We're stroppy, we're aggressive, we'll take over the world.* Then there

19

are the Bob Marley, Michael Jackson and Pink Floyd songs that dominated our school radio station in 1980, the year we started high school. Also on the playlist are songs from bands we saw play live in the years that followed – Split Enz, Ultravox, Unrestful Movements, Siouxsie and the Banshees – and songs from the albums we played over and over, by Duran Duran, the Cure and Joy Division.

Then there's Bob Dylan, Amy Grant and the Resurrection Band, with their lyrics about angels, 'the Lord', 'the blood of Jesus'. Music from that intense two-year period we shared. The time when evangelists were preaching about the End Times, and we were convinced the Pope was the Antichrist and that EFTPOS cards were the beginning of the 666 system. The time when churches were starting petitions against homosexual law reform and the ratification of the United Nations Convention on the Elimination of All Forms of Discrimination Against Women. The time when we didn't know which would come first – nuclear annihilation or the Second Coming of Jesus.

Sometimes Maz and I refer to those two years with a quiet 'Faaaaark' or a shake of our heads, but we've never talked much about it. Now, decades later, I want to talk about it, to see if we can understand our born-again Christian phase from the safety of nearly 40 years' perspective. Things that are happening in the world now – the conspiracy theories about COVID-19, the rise of the alt-right, the eroding of women's rights in the United States – make it suddenly feel very relevant. And I've noticed that the people most likely to believe conspiracy theories are often disillusioned, disenfranchised, disadvantaged in some way. So what was it that made me and Maz – two smart, middle-class girls in 1980s New Zealand – so vulnerable to those born-again beliefs? Why did we have such a need to belong to something?

The road between Nelson and Westport is not bad, Maz says. 'For the population density, roads in New Zealand are incredible. We spend a lot of money on roads per capita.' I'm not particularly interested but listen as she continues. Maz has been working on roads a few years now. She's currently project manager for the Auckland Skypath Project, which as we're driving seems to have hit a blip after the plan to add a cycle lane to the Harbour Bridge was reported as being unfeasible. The news was met with outrage from cyclists. As we drive through a series of small settlements – Belgrove, Motupiko, Korere – the radio delivers updates on the 'Liberate the Lane' activists, who are staging a mass cycle ride across the Harbour Bridge. There's also news of wild weather hitting the east coast of the South Island, where prolonged heavy rain and king tides are causing flooding around Christchurch and Banks Peninsula.

When she's finished lecturing me about roads, Maz tells me about the people we're going to stay with in Westport. Emily makes gourmet pies and sells them to businesses along the coast. Tom works for Stockton Mine, the biggest employer in the region. West Coast people have a reputation for being rugged, self-reliant, individualistic – and hospitable. I've never lived on the West Coast but I'm descended from coasters. Three generations of them lived here – my great-great-grandparents, my great-grandparents, and my maternal grandmother – before dispersing to other parts of the country. For me, 'the Coast' feels like the closest thing to an ancestral home that I have, with family stories referencing Hokitika, Māori Creek, Lake Kaniere, Kokatahi. Maz has Coast connections too. Her gold miner great-grandfather, who was born in England, went to the School of Mines in Reefton, and started a family on the coast.

Across the Hope Saddle we see snowy mountains shrouded in mist and we start to head southwest. This is Ngāi Tahu country, and the road follows the Kawatiri Trail, an old pounamu route that runs beside what the English settlers named the Buller River. It's cold – seven degrees – and wet. Little waterfalls flow out of the bush, down the road-cutting and onto the verge.

'The G-forces were getting a bit strong on that corner,' I say, clutching the grab handle above my seat. I'm determined not to let anxiety get in the way of our trip, but Maz is driving like an F1 driver.

She's talking about heading back to Westport, where she lived for four years, and is feeling nostalgic. 'I always fantasise about living in these small towns, but you know that within five minutes everyone's going to know your business. And it's a hotbed of bed-hopping.' Before Westport, Maz was in Australia for 19 years, spending time in Mudgee, Mt Isa, Kalgoorlie.

The Buller River is brown and fast, in full flood. As we drive, Maz tells me about outdoor antics from her Australian mining days, like when she went canoeing in a flooded Leichhardt River at Mt Isa. She was working at Mt Isa Mine, a shift boss working nights, and decided to go down the river in a Canadian canoe with one of her shift crew, Richard.

'It's normally a muddy trickle, but when it pisses down with rain in the outback it becomes a raging torrent,' she says. 'We got bicycle helmets, life jackets and our paddles and jumped in. We lasted about a minute.' She fell out, lost her mate and the canoe, and was smashed into a gum tree in the middle of the river. 'I didn't know where Richard was, and I couldn't see the canoe. I climbed up to a higher limb, then these guys heard me shouting and eventually got me out of the river with a rope.'

'Was that a risky situation?'

'It was fucking insane.' She can hardly speak, she's laughing so much. 'We should never have attempted such a ludicrous thing, canoeing down a raging river, water halfway up 20-metre gum trees, with waves, fucken logs and everything being swept down the river.'

Richard was injured, trapped between a tree and the canoe. He didn't make it to work that evening.

'What happened to the canoe?'

An hour or so later, she says, a mate who'd been downstream, standing on a bridge looking out for them, saw it float past, upside down.

By the time she's finished the story, we're both cry-laughing. It sounds dangerous as fuck, but it's the sort of thing that's exciting to have done. We've both done some things we're happy to have come out of alive.

'I look back now and shudder,' she says. 'We could have died.'

We have one stop planned before we hit Westport. Steve had told us to meet him at 1pm on the main street of Murchison, State Highway 6. We arrive early, so park up on a side street then go exploring. There's an op shop with the usual mix of second-hand clothes, records, books and knick-knacks, and a butcher with a handwritten sign in the window: *Home Kill Still Available.* In the local Four Square I stock up on fresh fruit and lollies while Maz tries on hats and sunglasses from a rotating stand. She raises her eyebrows at me. I shake my head at the bright pink plastic sunglasses but nod at a grey tweed trapper hat. It's no match for my West Coast possum fur trapper hat but a good look nonetheless. I snap a photo and we text it, without context, to our mothers, mine in Wellington and Maz's in Auckland.

At 1pm we look up the road for a Land Rover. Steve is easy enough to find – a tall, lean, grey-haired man, maybe a decade older than us, standing next to a mud-splattered dark green 4WD with a Natural Flames logo on the side. Below the image of mountains, bush and flames, it says 'visit New Zealand's extraordinary HOT spot – burning in Murchison since 1922'.

Two days from the start of winter, on a wet cold Sunday, we're not surprised to be Steve's only customers. 'Not many people would go to the flames on a day like this,' he says. Steve is already geared up in a black beanie and blue Columbia rainwear. During a phone call earlier in the week, I made it clear we were keen to go on the trip whatever the weather. I told him that between us we were trained in geology and mining engineering, and we weren't that fussed about the farm animals and nature aspects of the tour promoted on his website; we were interested in what was under the ground.

After getting us to sign some health and safety forms, Steve unfolds a large geological map and starts briefing us about the trip we're about to embark on. The map, from 1984, is like a riotous 70s abstract, each colour and texture combination depicting a different surface rock. The map covers a 270-square-kilometre area, with Murchison centre-west and the Tutaki State Forest in the east, and is dominated by the oranges and yellows of young sedimentary rocks. The youngest deposits – floodplain alluvium, glacial outwash gravels, and landslide deposits – follow the river lines. Outward from there are the oranges and peaches of older sandstones, mudstones, limestones, and what the map calls 'coaly layers'. Black dashed lines indicate faults, which mostly run northeast to southwest, the same orientation as the Southern Alps and the plate boundary that dictates much of the topography in this area.

Steve points to a river on the map. 'The Blackwater, where we're heading up, is above an anticline, which is –'

'We both know what an anticline is,' says Maz.

A cross section next to the main map shows the Blackwater anticline, a triangle of peach-coloured sediments deposited millions of years ago, pushing up through the Tainui Fault Zone, which at the surface is a narrow but heavily faulted strip of land along which the Blackwater River flows.

'This is where the oil comes up,' says Steve. While the river flows north, in the heavily fractured sandstones and mudstones below the river valley, oil and gas gather in porous rocks, some finding a way up through cracks to the surface, the oil pushed up under pressure, the lighter-than-air gas buoyant and compelled to reach for the surface. That's what we're here for.

⁓

We get into Steve's Land Rover and drive back along State Highway 6, then turn southeast down Mangles Valley Road. As we drive, Steve tells us about the prospectors who arrived in the 1880s to pan the valley's gravels for gold transported from the alpine ranges by Pleistocene glaciers tens of thousands of years ago. Next came the farmers, then the oil prospectors, attracted by methane seeps and rocks that smelled of oil. They drilled the first wildcat well, Murchison-1, in 1926, up the Mangles River. With a steam-driven oil rig fuelled by local timber, they reached a depth of 1245 metres but found only traces of gas and oil. The 1929 Murchison earthquake, a magnitude 7.8 quake with its epicentre just a few miles west of the township, killed 17 people and, according to Te Ara Encyclopedia, left the local area 'a shambles of fissures, landslides, floods, and destroyed roads, bridges, and buildings'. In the next attempt, in 1968, the Blackwater-1 well reached a depth of 637 metres and – I learn from a laminated news clipping Steve gives me – produced 'promising traces of oil and gas', though the manager of the company drilling the well said he was 'not exactly jumping with excitement' about the prospects. For the most recent attempt, in 1970, a 500-tonne rig from Argentina, with a 40-metre-high tower, drilled further up the Blackwater Valley, spudding in at the Tutaki sandstones and bottoming in 3131 metres down, at the top of some older sediments. 'Big party at the end of the drilling, lots of Argentines here,' says Steve. Despite all those past attempts at drilling for oil, the only reliable thing is the methane seep we're going to see, hidden deep in the forest, and never disturbed by drillers.

I sit in the back, leaning forward to listen to Steve and Maz talk West Coast language full of references to coal prices, pie shops, and who has the concession on which gold claim. Today the Mangles River is in high flow. 'The kayakers like it when it's flooding,' says Steve, and Maz and I giggle, thinking of her Mt Isa escapade.

'In summer, guys turn up with a suction dredge and jump in with wetsuits – they have a gold-mining claim on the river,' says Steve.

On the far side of the river is bush. On our side, the road is bordered by fenced-off farmland, with the occasional letterbox. The landscape is wild and bluffy. In winter, says Steve, feral pigs and goats emerge from the hills and dig up the fields. Big rocks came down here after the 1929 earthquake, he says, prompting Maz to talk about her engineering work on the Lyttelton to Sumner road in Christchurch. Damage caused by the 2010–2011 Christchurch earthquake sequence – rockfalls, debris flows, collapsed retaining walls – closed the road for eight years. Maz led the contracting team on a massive remediation project, and is clearly proud of the work.

I'm happy in the back, just listening, but when Steve admires my hat I boast that it's real possum fur.

'That's not made by that guy down South Westland, is it?'

I bought my hat a few years back, at Scott Base, Antarctica, from a field trainer who lived on the West Coast.

'Chris Long?' I say.

'Yep,' says Steve. It's a very West Coast thing that he can identify the maker of my possum-fur hat. Steve hasn't met Chris, but knows of him from his blog, Wild Kiwi Adventurer. 'He's a bit of a character.'

We ask how the border closures have impacted Steve's business. At Christmas there were fewer tourists than normal, he says, but late summer was quite good. 'Lots of Kiwis from the North Island.'

'The JAFAs are keeping you going,' says Maz.

Steve describes where we're going as 'up the Blackwater'. At some point the public road ends and we head south along a bumpy gravel road that follows the Blackwater River, then turn west onto a farm road. Bob looks after the farm, says Steve, which looks to be populated with ruminant animals – sheep, cattle, deer. 'He also traps possums. He has a possum plucker.'

Maz is contemplating her romantic prospects.

'Are there any wealthy single farmers around here you can introduce me to?'

'I'll have a think about that. What age group?'

'I dunno, maybe around 50 to 65? Depending on fitness?'

'There is a single bloke . . .' Steve begins.

I look out the window. In a patch of roadside grass encircled by a thin electric wire lies an enormous black bull. He watches us as we pass, thick rivulets of water running off his hide. Up the road a ways we pass a tiny wooden shack with five pairs of Red Band gumboots lined up outside. There are no lights on, no smoke coming from the chimney. Bob's not home.

'Was he the one you were thinking of?' asks Maz.

'Well, he is single.'

We drive on.

'It's getting wetter and wetter, you do realise that,' says Steve as the drizzle turns into a steady rain.

Soon there are high-fenced paddocks on either side of us. Behind the fences are the deer – Wapiti bull bred with Red hinds for 'a bigger animal', says Steve. They were grown for meat then, for eating. I imagine what it would be like to be born for the express purpose of being eaten. The next paddock is full of young deer – smaller creatures that Steve refers to as 'replacement deer'. They stop grazing and watch as we drive by.

'Very pretty. They're all ears and nose,' says Steve. 'You sometimes see wild fallow deer here along the track.'

We're now off-grid. We pass a house, with a micro hydro set-up generating power from a stream. Steve slows down as a farm dog – a huntaway – circles the car, barking, trying to herd us as we drive past.

The rain continues and Steve asks again if we are still keen. We assure him that a bit of rain isn't going to bother us. 'The only thing I'm concerned about is safety,' I say.

The road is bumpy and wet. My concern with safety leads Steve to make 'a bit of confession'. He doesn't have a spare tyre today; it's getting fixed. There are also a couple of river crossings that might be a bit hairy in the rain. A slippery log, stepping-stones that might be inundated if the stream is up.

I log the hazard, but I don't know Steve well enough to tell if he's underplaying the risk or making it sound more dramatic than it is.

The last bit of the drive is off-road, across a rolling green paddock surrounded by bush covered hills, and I bounce around in the back of the Land Rover. We pull up next to a high wire fence on the other side of which leans a row of coloured walking sticks. Steve opens the back of the vehicle and presents us with a selection of waterproof parkas and overtrousers. After kitting up – me in retired yellow ambulance gear, Maz in blue waterproofs – we each grab a stick and follow Steve along a faint path into the trees.

The forest is dominated by beech. The bush is close, dripping wet trees, ferns, fungi and mosses, and everywhere I look I see decay. White bracket fungi jut out from some of the trunks, while tiny, pale cups dot the dead wood below. The red beech trunks are black with sooty mould – a type of fungus that grows on the honeydew excreted from the anuses of scale insects – which drips onto the earth below. The mould gets us talking about COVID-19. There is recent news from India about a life-threatening 'black fungus' infecting people who've recently recovered from the virus. 'A nightmare within a pandemic' is how one doctor described it.

We walk with heads down in the rain, enjoying the closeness of the forest. The rough track is barely perceptible, softened and blurred by a layer of tiny beech leaves, yellow and brown, with shallow tree roots snaking across the ground, the smaller ones creating trip hazards, the larger moss-covered ones interlocking to form complex caves and warrens. We see a dinosaur skull, elven bunkers, Shinto shrines. It feels like the sort of place where you'd find elves, kami, patupaiarehe.

'There's an archway to another world,' says Maz, pointing

to a high-arched and moss-covered root. 'This whole place feels like another world compared to downtown Auckland where I was at 6am this morning.'

'Pre-COVID, I remember bringing a lady up here only 48 hours after she was in the middle of Amsterdam. Imagine that. Straight from Amsterdam to a place like this,' says Steve.

As we walk, Steve tells us about the birdlife – this forest is home to robins, pīwakawaka, riroriro. But it's been a bad season for birds, he says, because of the mega-mast in 2019, during which an unusual bounty of beech seeds provided food for mice and rats. 'What's good for the rats is bad news for the birds.'

We notice a large, dead, red beech. I look around, naturally hypervigilant and thinking of the Murchison earthquake. 'I reckon if it really shook, a lot of dead wood would come down.'

'I've never thought of that before,' says Steve.

'Yeah, see how many dead trees there are around here.'

Maz shrieks. 'We'd have to crawl into one of those elven bunkers.'

'I think the best bet –' I begin.

'I would just drop, cover and hold right here,' says Maz, pointing to the base of a tree where a root system has created a cosy cave-like hole.

I look up. 'That's a dead tree, Maz.'

We start giggling. 'And then it would tip over and I'd be catapulted across the river,' she says, eyes gleaming.

'I wonder whether standing under a nice young tree would be better?' offers Steve.

'Something with a bit of bounce in it,' I say. We agree that's the best approach, then carry on along the trail.

We get to talking about what we'd eat if we were stranded in the bush. First, Steve points out some coprosma, a small scrubby bush that grows orange, red or even blue berries in autumn. They were a traditional food, he explains – mingimingi – but I've not heard of them. 'You'd have to eat hundreds to make it a meal,' he says. I recognise the horopito bushes by their pale green leaves with pinky-red blotches. I wonder if the heat is in the red spots, so I test my hypothesis by ripping a green leaf off a bush, putting it into my mouth, and chewing. Hot! Then I try a red one. Hotter! I spit out the leaves – I guess they're not really a snack food – and load a handful of lollies into my mouth.

It was Maz who insisted we stock up on lollies. She was diagnosed with post-viral syndrome in 2003, told by her doctor to be 'stoic' and just deal with it, but lately she's been reading a lot about the syndrome. The things that seem to help are stress avoidance – 'a real challenge' – plenty of rest, red wine, and an array of expensive supplements containing nootropics and antioxidants. Her latest theory, which she's testing today, is that glucose is another winner. In an attempt to avoid the two-hour walk sending her into a relapse, she's downing handfuls of lollies. She's told me she's desperate and will try anything. She hasn't done a walk this long in years, and I know she doesn't want to let me down. I'm eating lollies in solidarity, and we're taking it easy, letting her set the pace, but she's doing fine.

Steve points out a wasp bait station and we notice an enormous kahikatea. With its massive height and huge trunk it stands out against the skinnier beech trees. 'He survived,' says Steve. 'They cut them down, for butter boxes, in the 1950s.'

'The 50s have a lot to answer for,' says Maz.

The land is now conservation land, the trees can't be touched, and young kahikatea are starting to grow back, but it will take hundreds of years to grow another giant like this.

'Let's just hope they don't find any gold here,' says Maz.

'We know the oil will stay in the ground, but gold's still fair game.' Steve asks if there's any chance of gold being found up at Stockton Plateau, where we're headed tomorrow, but Maz says it's not the right kind of deposit. 'There have been some whispers, though, that some of the overburden might contain rare earths, but maybe nothing economic. I doubt there will be anything there, but you never know.'

'You never know when things might become economic,' I add.

Steve and Maz are talking about oil, coal, gold, rare earths, but I'm thinking about uranium, which I wrote about in a book ten years ago. Back in the 1950s and 60s a series of individual prospectors and then prospecting companies searched for and found uranium in the Buller and West Coast areas. It was the dawn of the nuclear age, and New Zealand was looking for a supply of uranium to fuel the nuclear power plants that would provide electricity to our growing population. Companies from Australia and Germany, supported by funding from the New Zealand government, spent years searching but the deposits found were never rich enough to make extraction economically viable. Then, in the 1980s, following our nuclear free policy, it became illegal to even prospect for uranium. But people are starting to talk about nuclear power again today, as a low-carbon electricity-generating option, and I mention that some of the gold tailings on the black sand beaches of the West Coast are rich in thorium, a mineral used to fuel a new generation of cleaner, safer nuclear reactors.

⁓

About an hour into the walk, after steady rain and a couple of stream crossings, our oversized waterproofs are wet and dripping.

Steve tells us we're nearly there. The track heads away from the stream and up a gentle rise. We climb some wet wooden stairs, then grasp hand ropes to haul ourselves up a steep root-lined path. Along a short flat section of track and up a wooden ladder, we find ourselves on a rise, in a little clearing backed by tall beech trees, surrounded by lush ferns and grasses. At the high point of the clearing we find a little alcove lined with wooden benches. Steve takes off his pack, indicates a space for us to sit, and busies himself with some metal grates and kitchenware. We drop our bags and gape at the hollow below us. In a rocky pit set into an earthen bank, a patch of dirt about two metres square is on fire. Yellow, orange and white flames cavort against the burnished earth. Over the sound of gentle rain is the bubbling sound of gas and the roar of the fire.

The story Steve tells us is that back in 1922, a couple of hunters found the methane seep – they heard and smelt it – and one of them set it alight. It's been burning pretty much ever since. The location of the flames moves, says Steve. He trims back the vegetation over a fairly wide area, to ensure nothing is in reach. 'If a tree fell right in the bang smack middle of it, in a dry summer, I'd be concerned.' In all the years that Steve's been bringing visitors here, he's only put the flames out once, when there were wildfires in the area and he didn't want to risk adding to the problem. He relit them once the rains returned. I'm worried about the burning fossil fuels, but Steve reminds me that the combustion product – the carbon dioxide released by the burn – is not as harmful as the methane gas that would otherwise be finding its way into the atmosphere.

Now that we've stopped walking, we can feel the cold. Maz and I warm ourselves by the flames, which flicker a foot or so above the ground, then sit on the benches and watch Steve work, ferns damp behind our backs. Around the camp there are metal implements hanging from branches – a piece of bent reinforcing iron, a cow bell, a grill and a large pan – giving things a bit of a macabre feel, like there's something going on here that's painful, ritualistic. I wouldn't want to come across this place if I was alone in the woods. Steve, who is cheerful and reassuring, boils water in a billy and heats an iron griddle over the fire. Before long we're drinking billy tea – smoky Lapsang Souchong – and eating pikelets drizzled with Tutaki honey made by bees that forage on honeydew in the beech forest. I exclaim to Steve that they are the best pikelets I've ever eaten, deliciously light and fluffy. He uses the same recipe as me, Edmonds, but reckons the shaking and mixing the batter gets on the truck and in his backpack aerates the mixture and makes for a superior pikelet.

'Are there any more places like this in New Zealand?' I ask Steve.

'No, it's quite unique.'

While this is the only one in Aotearoa, there are other flaming methane seeps around the world – Turkey has the Flames of Chimaera, Turkmenistan has the Gates of Hell – but none of these other fires are in a forest.

When we've finished eating, we sit staring into the fire, like humans have for eons before us. I waver between seeing the flames as a comforting source of heat and as a gateway to some netherworld inferno, a lake of fire. My science brain knows what's happening here – methane from sediments deep in the earth has made its way to the surface and is combusting, releasing carbon dioxide and hydrogen gas – but it also feels somehow sinister, diabolical.

Maz turns her gaze from the everlasting flame and hits me with one of her mischievous grins. 'Where's Satan?'

3

Salvation

In the 1980s, Maz and I lived in a world that already felt damned. In this New Right era of Reaganomics and Thatcherism, we spent a lot of energy raging about apartheid, yuppies and the nuclear arms race, along with the racist patriarchy that generations before us had endured. So it wasn't a fear of damnation that led us to Christianity. Christianity offered us an out, a free pass to a world of purpose and meaning and angels watching over us. Our fear of damnation didn't come until after we were saved.

It was 1983 when some of our friends who'd been brought up Catholic started going to charismatic Christian youth groups. Other friends went to crusades in Wellington where they were saved. These born-again Christians were smiley and friendly, and it made our angsty schoolgirl selves feel happy to spend time with them, sitting around bonfires on the beach, talking about the meaning of life while someone played Bob Dylan songs on guitar. Some of the boys were really funny and made us laugh and forget that everything was fucked.

While there was something exciting and enticing about these happy groups of born-again Christians, it made no sense. These were the people our mothers had warned us about.

They'd told us about the Billy Graham crusades in 1959, had seen people 'born again', and knew enough about evangelical Christianity to tell us to watch out for 'the Pentecostals' or 'happy clappers'. We'd grown up in families without religion, though we'd independently gravitated towards local Sunday school and Bible classes. Maz had a nuclear family, two parents and a younger brother. For my teenage years I lived as part of a blended family, with a mother, a sister, a stepfather and two step siblings. When we asked our mothers what religion we were they laughed and said we were pagan. But now we found ourselves drawn towards this opportunity for salvation, this promise of a better world, this *certainty*.

We started praying, 'Dear God, please help me to believe in you.' If He was real, He would give us faith, and if He wasn't, then what was the harm in praying? When we told our Christian friends we had been praying they said it didn't matter that we had doubts; Jesus would take care of that, and they led us through the Sinner's Prayer and we invited Jesus into our hearts.

Dear Lord Jesus, I know that I am a sinner, and I ask for Your forgiveness. I believe You died for my sins and rose from the dead. I turn from my sins and invite You to come into my heart and life. I want to trust and follow You as my Lord and Saviour.

Most of the born-again Christians we knew lived in the city, so we found our own church in the seaside suburb near where Maz lived, an Assembly of God with a small congregation of worshippers who every week greeted us with delight and enthusiasm, like they were perpetually surprised to see us. One week, the pastor made us stand on our Bibles to make the point that the Bible was now the foundation of our lives.

We felt uncomfortable about standing on a book, but we got the point.

We had never done things by halves, and once we had accepted Jesus Christ as Lord, we hit it hard. We ripped the posters of the Cure and the Sex Pistols from our bedroom walls. We stopped defiling our bodies with alcohol, cigarettes and pot. At school, we started smiling at our teachers and working hard to make use of the gifts given to us by God. In the evenings we did our homework, watched wholesome TV programmes and read our Bibles. We prayed for our little brother and sister – that they would get saved too. We prayed that our verrucas would go away.

On Fridays, we looked in the newspaper to see if there were any weekend crusades on. At our first crusade there was an altar call, and we walked to the front of the auditorium where the visiting evangelist stretched out his hand towards our foreheads. We fell backwards – slain in the Spirit – into the arms of tall, strong young men, who lowered us gently to the floor while we closed our eyes and praised God. The friends we'd invited, hoping they'd be saved too, arrived late, pissed and laughing at us.

At the crusade, we learned that Christianity involved more than praying and believing in God. Because Jesus was coming back, we had to be vigilant that our words and deeds were pure. If He came back *right now*, would we be ashamed or proud of what we were doing? We had to break up with our boyfriends, who were also best friends. When we told them, they smoked a joint, drank a flagon of beer, then left. They told people we were bitches because we had gone Christian on them, so we prayed for them.

At another crusade, the pastor was casting out demons. You knew people were demon-possessed because when the pastor commanded the demons, in Jesus's name, the demons got angry and defiant and would manifest. Sometimes they would shout, swear and lash out. The pastor cast out spirits of anger, disobedience, homosexuality. When he called on the spirits of masturbation to depart we watched, mortified, as a girl our age walked forward, sobbing loudly and shaking her hands furiously to cast out the evil spirit.

For a long time, though, we were lukewarm Christians, because as well as going to church and crusades we still hung out with our old friends. Sometimes they would come to our houses to drink their beer and smoke their joints because we had cool parents. Some of the boys liked us but nothing could happen because they were worldly and we were Christians. We had a lot of pent-up energy back then. Sometimes when the boys came around we would hit them with cushions. Sometimes we had to let off steam by going for a run around the block.

On Sundays, we would go to church and repent for associating with sinners. After a while, though, the people at church told us we had one foot in the Church and the other in the World. 'Be ye not unequally yoked together with unbelievers,' said one of the elders, so to show we were serious we took our worldly records to the top of the bank behind Maz's house and Frisbeed them off the cliff, watching the wind catch the black vinyl discs and then dash them into the trees below. We made a fire and burned our teenage diaries, our horoscope guides, our Hare Krishna recipe books. We stopped hanging out with our old friends altogether.

Earth was a spiritual battlefield, but with enough faith, we could heal the sick, cast out demons, perform miracles. With a newfound sense of clarity about our lives, we started overflowing with feelings of happiness, peace and joy.

One Sunday after the service, the whole congregation got into a minibus and we drove out of town, along windy gravel roads into the bush. Maz and I were ready. On the stony banks of a fast-flowing river we slipped white robes over our T-shirts and shorts then waded into the water. We crossed our arms over our chests and the bearded pastor dipped each of us quickly backwards until we were fully submerged. When we came up, everyone clapped and smiled and hugged us. We ate a picnic lunch beside the river while our hair and clothes dried in the sun.

Now that we were baptised Christians, our mothers persecuted us even more. They would burst into song when we entered the room – 'Onward Christian soldiers!' – followed by a cackling laugh. Sometimes we wondered if they were demon-possessed.

The elders at church said we should meet some young people from their sister churches, so one weekend we caught the train up the coast to a Christian camp. On the first evening, we ate toasted white bread for supper, then listened to some of the young married couples talk about their lives in Christ – about the importance of fidelity, prayer and one hour's Bible study a day. In bed that night, we whispered to each other from our bunks, marvelling at the air of happiness, purity, and cleanliness around us, compared to the darkness and filth of our previous lives in Wellington. We prayed to the Lord to help us meet some new Christian friends.

The next day, God answered our prayers. We met boys with names from the New Testament, like Paul and Simon and Mark, who'd been Catholics before they were saved. We used to think Catholics were Christians too, but we were told that was just part of Satan's deception. When we got to know the boys better, they would tell us, in quiet, reverent voices, that there was a good chance the Pope was the Antichrist but that this was not something they could say at home, in front of their parents. They were older than us and had already left school to work as exhaust fitters, warehouse managers, window cleaners. They sat next to us at mealtimes and commented on how different we were, not just from the other girls at camp, but from each other – one quiet and shy, one friendly and outgoing, they said. One cute boy with a cheeky grin, Bruce, drew attention to the way we dressed, saying our unfeminine clothes and hairstyles were signs of a rebellious spirit. Most girls, he said, wore clothes from mainstream fashion shops, and had shoulder length or long hair. We'd clearly gone out of our way to look different. Why were we rebelling against femininity? But he winked when he said it, so we didn't know if he was being serious, or just saying something that other Christians thought.

After dinner and Personal Devotions, we walked over an arched wooden bridge to a small island and sat around a bonfire to sing songs of praise to the Lord. The moon was full and it shone out from behind a cabbage tree and made a reflection on the still lake. We huddled in our coats and smiled in the moonlight.

The boy playing guitar finished strumming 'Oh Lord, You're Beautiful', then called out, 'Any requests?'

"'Jesus is Coming!'" said someone in a loud voice.

It was a song we knew from church, but everyone acted like he meant the words literally and people started jumping up, looking across the lake, shouting, 'Where? Where?' and laughing. We all imagined how good it would feel when Jesus really did come back and our souls were uplifted. By 3am most people had gone to bed, but Maz and I and two of the boys we'd sat with at dinner dragged kayaks down to the water and quietly paddled around the margins of the lake, while frogs croaked in the reeds.

When we left camp, the tall friendly boys with the love of Jesus in their eyes told us we should come to the Upper Room, a Friday night youth group at their church.

4

Coal seams

It's dark when Maz and I arrive in Westport. The air has a smell I recognise as coal smoke, but it's not the cosy, familiar smell of my grandparents' house in Christchurch; it's rancid and eggy.

'A lot of coal in Westport is high in sulphur,' says Maz. 'That's why it smells gross.' Sulphur dioxide, I think. Stinky, toxic, corrosive. A nasty pollutant released alongside the greenhouse gas, carbon dioxide, which is emitted whatever kind of coal is burned.

We drive down the main street and around the shopping centre until we find a liquor store. I choose a local gin, Little Biddy, made nearby in Reefton. The guy behind the counter recommends it.

'You could drink to here,' he says, indicating about a third of the way up the bottle, 'and not feel it the next day.' I'm sold. Back in the car, we navigate through dark suburban streets before pulling into the driveway of an art deco house set back from the road, with a luxuriant front garden of succulents and palms.

Inside, Emily is cooking. She apologises for the kitchen, says it's mid-renovation. But it looks great to me, with open industrial shelving displaying cookware, preserves, bowls of onions and garlic. I make gin-and-tonics while Maz and Emily talk. When Tom arrives, Maz comments on how much he's changed since she last saw him. I glean that perhaps he has a new hairstyle, gained a bit of muscle?

'I haven't found God, but I have found Lycra,' he says with a laugh. He's just back from a bike ride.

When dinner is ready, we sit around the table while Emily serves. Her business, the West Coast Pie Company, makes pies with wild meat – pork, venison, hare, rabbit. She's thinking about adding tahr and wallaby. 'I want to bring back the West Coast three-course meal – a pie and two beers,' she once said in an interview. But there are no pies on the table tonight. Given Maz's pre-trip response to my largely vegetarian diet – 'There are no beans and quinoa on the Coast, we might as well cancel the whole trip!' – I've decided to ease up on things if someone else is cooking, especially a chef, and I asked Maz to tell Emily that my only dietary restriction is red meat. Dinner is roast pork. I forgot that pork is generally referred to as white meat. Like human meat, I think. I guess I should have said that I don't eat fellow mammals. But there is also a lentil and pumpkin bake, homemade flatbread, and a yoghurt garlic sauce. I serve myself a generous helping of lentil and pumpkin, flatbread and yoghurt, then take a tiny piece of pork crackling. It's important not to be dogmatic, rigid, black-and-white about things, I tell myself as I crunch down on the roasted pig skin.

After dinner, we sit around the table with a bottle of red wine. There's a fireplace, and the open-plan kitchen-dining-living area is toasty warm. I ask what's providing the heat. Tonight, it's wood, says Tom. But 'when it's super freaking cold' they burn coal, for its 'amazing heat'. Maz looks at me and grins. She likes winding me up about coal, and she's enjoying the shock on my face, my realisation that people outside my circle of Wellington greenies would even consider such a thing. I grew up with coal fires – my grandparents had coal delivered by truck to a shed at the back of their house, and sometimes it was my job to fill the coal scuttle that sat on the hearth – but it's been a long time since I've sat by a coal fire or even seen a bag of coal.

Once she's caught up on the Westport gossip, Maz tells Emily and Tom about her recent separation from her husband. It's been a tough year, but she's good, she tells them. 'I'm reading books, losing weight, drinking less.'

They start talking about single men in the area.

'Fred is single, he's good-looking,' says Tom.

'But isn't he gay?' asks Emily.

'There's rumours.'

Emily offers 'a good-looking, single, born-again Christian' she knows and Maz and I laugh. Maybe not. What about a farmer? Westport is now surrounded by dairy farms. There are bound to be some farmers in the area who are single, a bit lonely.

But Maz says she doesn't want a farmer, she wants a sailor, so she can go on ocean adventures. I know she's not serious though, not about any of it. It's way too soon. She's still grieving the end of her marriage, on good terms with her ex. I think she's just testing saying it out loud, entertaining the possibility that maybe one day there will be someone else.

After dinner, Maz and Tom – who works at Stockton Mine – start talking shop. It's a language I don't understand: overburden ratios, Schedule 2 wetlands, potash from Belarus, industry pay rates.

After our earlier encounter with flaming methane – a gas

that miners call firedamp – I want to ask Tom about mine fires. I've heard that up at Stockton there is some coal, still under the ground, that is on fire. It's true, he says. There's a section of the mine that has been slow burning for a hundred years or so, spewing out carbon monoxide, hydrogen sulphide, sulphur dioxide.

Tom tells us that underground mining on the Stockton Plateau started in the 1890s. Open cut mining started eighty years later, in the 1970s, but they couldn't get on top of the mine fires.

'I've put out three fires and have probably got five more to go. The basic process is you use drilling to determine where the fire is, then create a programme of blasting and dosing with water to put the fire out, then you've got three months to mine the coal out before it decides to reignite itself with latent heat.' He explains that there's a pocket of coal close to 'spon com' – spontaneous combustion – at Stockton, which has always been an issue.

Talk turns from mine fires to spontaneous human combustion. I worry much less about this phenomenon than I once did. Today, I worry about the fossil fuels we've been burning since the start of the industrial revolution, which have increased carbon dioxide levels in our atmosphere from 280 to 420 parts per million and climbing; trapping heat, warming the oceans and atmosphere, and causing climate disruptions all over the planet. There's a lot of talk about action on climate change, from governments, businesses, the Intergovernmental Panel on Climate Change, but we're still burning coal, we're still putting petrol in our cars, we're still cutting down forests to graze animals for meat, we're still wanting *more*, and global carbon emissions continue to rise. We need to make changes, and fast.

I ask Tom how he reconciles his understanding of climate change with working at a coal mine.

'I have no trouble with that at all,' he says. 'I sleep just fine at night.'

It's all about the quality of the coal and what it's used for, he says. I already know from Maz that the coal mined at Stockton is some of the best in the world. It's not used for heating homes, factories, hospitals, not used for generating electricity. These are all things that can be done with alternative energy sources. This is coking coal, metallurgical, exported for use in steel-making in Japan, India, China, South Africa and Brazil. It has high levels of vitrinite and low levels of ash, reducing the amount of coal required to produce a tonne of steel and therefore reducing the amount of carbon dioxide generated in the steel production. So is it somehow 'good coal'? After all, if the steel-making companies can't get Stockton coal, they'll have to burn inferior coal, which would release more carbon dioxide into the atmosphere.

It is good coal, says Tom, who gets annoyed at calls to 'keep the coal in the hole', to stop mining coal altogether. 'The youth have no idea where stuff comes from. This whole green revolution is hinged on finding rare earth elements that are probably up the road in Barrytown.' He's got a point. Green technologies like wind turbines, electric vehicles and low-consumption light bulbs all rely on rare earth elements with names like neodymium, praseodymium, samarium. And you need mines and processing plants to extract them from the ores in which they're found.

When Tom walks out of the room Maz lifts her top and shows Emily her boob tattoo – a delicate, fine-lined twist of vines and flowers that hides a transverse scar. If you look closely, there's a tiny skull in there, smirking in the face of mortality. The cancer diagnosis, the surgery, the marriage break-up – a lot has happened in Maz's life since she left the Coast.

We start bitching about work, about being ignored at meetings, talked down to, paid less than men, while Tom does the dishes. It seems like yesterday we were the underdogs, the young women fighting for a place in a male-dominated world. There were plenty of good men who supported us, sponsored us, recommended us – 'allowed' us to progress our careers. But there were other men who were condescending, patronising and generally assholes. And then the #MeToo movement made us acknowledge the things we'd put out of our minds as embarrassing or shameful and reframe them as sexual harassment and even sexual assault.

Before we head off to bed, we discuss our plans to go north the next day, through Waimangaroa, Granity, Hector, place names I know from my mother's stories.

'Lose your sanity in Granity,' says Maz.

I used to think of this part of the country as a sort of paradise. My teenage plan, when I found things too overwhelming, was to train as a GP (a useful skill for when it all went down), marry my schoolfriend Adam, and live on the West Coast, growing my own vegetables, being self-sufficient. It's not quite the life I'd want now. In Wellington, I have the best of both worlds, with a garden large enough to sit in in the sun, grow vegetables and fruit trees, find a place to hide. I also know that, alongside the stunning scenery and hospitable locals, the Coast has social and economic problems – unemployment, uneven access to healthcare, alcohol abuse – that have been exacerbated by the border closures and drop in tourism.

⁓

'As a bridge engineer, I can tell you my expert opinion,' says Maz. 'That bridge is fucked.'

We've woken to news of flood evacuations and road closures across the ranges on the east coast. We watch video clips of widespread flooding and damage, including to the Porter River Bridge on State Highway 73, the main route back to the east coast. That's the road we're hoping to take a week from now, when it's time to drive to Christchurch.

We start talking about what would happen if we got stranded on the West Coast, cut off from the rest of the country. We both find the prospect quite appealing. If I didn't have my family at home in Wellington, I'd be totally game. The year of closed borders and lockdowns that made our lives so much smaller, gave us permission to do less, to focus on home, garden, family, suited me. But the flipside was a serious Twitter habit, doomscrolling about the pandemic, climate change, the batshit crazy conspiracy theories. I quite liked the idea of being stuck in a small community, focusing on essentials, *surviving*, but I'd want to be with my family. And maybe the peacefulness would only last as long as the internet was down, the electricity off, for as long as I could block out all the crazy in the world.

I'm out of my sleeping bag and straight into yesterday's clothes. This West Coast roadie feels refreshingly like a field trip, or a tramp. We're back to basics: food, shelter, transport.

Emily waves us off, jealous about our mine tour. She can't come – she has 600 pies to make.

Our first stop is the main street of Westport, where we pull up outside a café. Maz strides in ahead of me, a wide smile on her face, and greets a man behind the counter.

After Maz tells Jay what she's been doing since leaving Westport, they start catching up on news about people with names like Fish, Roger, Swifty. I hang back, then place my KeepCup on the counter and am introduced as Maz's BFF. I'm happy listening, though, and feel reluctant to join the conversation.

Before we head out of town with our coffees, Maz takes a quick detour past her old place. It's a mid-century bungalow next to a new house her husband built. I stayed in the bungalow with my family one night, two parents and three kids in sleeping bags on the floor of the partly renovated house. Our entire South Island trip was in the wake of Cyclone Ita and we had slept well after battling our way from Christchurch to the Coast through flooded roads and torrential rain. Maz and her then-husband moved out, over the hill to Christchurch, seven years ago. And from there to Auckland.

Now the highway takes us north along the thin strip of land between the Tasman Sea and the Denniston Plateau. The road is empty, lined with fences beyond which lie grassy paddocks, harakeke, grazing pūkeko. It runs parallel to the railway line that takes the West Coast coal across the ranges. Above us the skies are clear, the bush-covered mountains ahead of us dark silhouettes against the morning light.

As Maz drives, we listen to news updates about the latest storm. More roads are closed, more bridges are down. 'Fulton Hogan and Downer will be rubbing their hands in glee,' says Maz.

Then the news turns to COVID-19 and the outbreaks in Australia and Fiji and we quieten down. Maz has just come out of a series of Auckland lockdowns and we're acutely aware that we're lucky we can travel. Less than three months after our road-trip we'll both be at home again, subject to another strict nationwide lockdown following detection of community transmission of the Delta variant of the virus.

At Granity, we turn right, cross the railway line and head up the hill to Stockton, where Maz used to work. The job that brought her back from Australia – though being closer to her aging parents was also part of the pull – was general manager of a small coal-mining start-up working the Cascade Mine, now part of the Stockton complex, and looking to develop new leases up on the Denniston Plateau. After that she worked at Stockton as a consultant technical services manager. Around this time, an underground explosion killed 29 miners at the Pike River Mine, inland and a couple of hours' drive south, sending shockwaves throughout the industry and leading to a new industry-wide emphasis on training, health and safety.

As we drive up the hill, Maz tells me about Stockton mine.

'Back in the day, they did a heap of underground mining here using bord and pillar techniques, which is basically a criss-cross mining pattern.'

Starting around 1900, underground miners extracted as much of the coal as they could, leaving pillars in place to stop the ground collapsing. Today's miners are using open cut techniques to try and extract all those pillars and any other remnant coal. But the voids where the coal was extracted in the past are gassy, treacherous, and some of them are on fire.

'So it's quite dangerous. You've got to watch out for ground collapse. You have to do highly sensitive mapping over the void areas, plus you've got to deal with really dangerous toxic gases. It makes the mining very technically challenging – but fucken interesting.'

It doesn't fit with my mental image of mining, which involves men with pickaxes rather than machines.

'That was 100 years ago,' says Maz.

That's still how it is in my head though. I think of black-faced miners walking out of a shaft after a day underground. Miner's strikes. Politics. Welsh choirs. *The Denniston Rose.*

We meet Tom in the carpark, where we pull in alongside a row of white Toyota Hilux with Bathurst decals. Tom takes us into a prefab office and introduces us to Merv, who's going to take us through our induction. First there's a form to fill out 'in case there's an emergency and we have to contact your next of kin', says Merv. While we're filling out forms, Merv offers Tom some of his breakfast. 'You can sample Russell's soup if you want, mate. I don't know what he's put in there – eye of bat, leg of toad or something.' Tom passes. 'Any word on Canterbury, mate?' Merv asks.

Maz chips in. 'It's still there – barely. Still raining.'

Now there's a briefing video and another form: we need to answer questions to show we've been paying attention. In the video a robotic female voice tells us that Stockton Mine is a joint venture between Bathurst Resources and Talley's. We must obey all signs, alarms and barriers. Follow the instructions of our escort. We must not come to the mine affected by drugs or alcohol; nor bring firearms, non-approved explosives, alcohol, illicit drugs or animals on site; nor enter or pass beyond fenced or barricaded areas. *Breaches of the mine site rules may result in you being requested to leave the site.*

The video tells us that Stockton Mine is the largest coal-mining operation in New Zealand, producing 1.5 million tonnes of coking coal a year, which is exported for use in the making of steel. There are 250 full-time employees and contractors at the mine. There are many hazards associated with open-cast coal mines, and we'll need to be wary of ground and slope instability, explosives and blasting, historical underground fires, large heavy mobile equipment and more.

It reminds me of arriving at Scott Base, Antarctica. The special clothing, the forms, the hyper-focus on health and safety – and the very real danger if you go walkabout without heeding the safety advice. *Drug and alcohol testing is conducted at Stockton Mine using both saliva and urine methods. As a visitor you may be randomly selected to provide a drug and alcohol sample to confirm you are not impaired.* This inspires some teenage outrage in me. Fuck that! I'm not under the influence of anything but coffee, but I tell Maz I'd leave the site on principle if anyone asks me for a 'sample'.

Maz laughs and tells me to get over it. 'If you want to be around heavy equipment and explosives, you need to know everyone is drug-free.'

Following our induction Merv issues us with hard hats and hi-vis vests. Tom checks out our footwear – we're already wearing heavy boots – and we're good to go. I sit in the back of the 4WD while Maz takes the passenger seat. Tom starts driving us up the hill on pinkish gravel roads.

Tom first came to Stockton as a geotechnical engineer, and is now technical services manager, with a team of 13. He's been here 18 years. 'I'm a lifer,' he says.

As we drive, Tom gives me a lesson on how the coal gets off the hill. He points out a structure on the skyline. Coal extracted from the mine is transported up to the ROM pad, to be sorted and blended into different mixes to suit the mine's clients. Geologically, there's one big coal seam, but the coal is extracted from different areas that are all part of the Stockton operation – Millerton, Cypress, A Drive, Rockies. Most of the coal comes from the big Millerton pit, but clients might want that blended with 'a sprinkle' of coal from Rockies or Cypress. 'So we'll take a bit of this coal, and a little bit of coal from somewhere else, and we put it together and we bake a cake, to whatever recipe the customer wants,' Tom says. Once blended, the coal travels down the hill on a conveyor, then is loaded into trucks and hauled another seven kilometres downhill, where it's loaded into a series of bins to travel down the aerial ropeway to Ngakawau. From there it gets put into the trains to Lyttelton. 'The trucks go non-stop,' says Tom. The route is earmarked for an electric truck system.

We've arrived. Tom has driven us to one of the highest points on the plateau, a gravelly platform overlooking a giant hole in the ground called the Mangatini Sump. Below the sparsely vegetated slopes, and surrounded by a dirt road, is a wide ovoid body of water, steely grey in the weak morning sunlight. It doesn't look like much, but it's a bit of a feat of engineering, created through a programme of drilling and blasting, followed by excavation and dumping of 1.6 million cubic metres of rock.

Most of the mine's dirty water – including the water that's been used to dampen down coal fires or wash the coal, along with contaminated rainwater runoff – is fed through a series of drains from around the mine and pumped up here to the sump. Over time, the sediments fall to the bottom of the sump, leaving a surface layer of clean water.

Next stop is back down the road, at one of the main pits. We stomp our boots up a pile of pinky-red earth to a sign saying *AUTHORISED PERSONNEL ONLY* and look into the pit. It's huge, hundreds of metres across. Tom explains. At the south end of the pit, the slope has been worked into a series of terraces stepping down to a dark, flat pool of water. At the northern end of the pit, a wide gravel road leads to a platform covered in dark grey rubble. On the very edge of the platform, above a 20-metre-high cliff, a drill rig stands next to a set of orange cones laid out in a grid. The rig is drilling a series of holes, nearby trucks loaded with explosives to pump down the holes. The goal is to blast off the sandstone cap rock, explode it into smithereens, and haul it away until the coal seam is exposed below. One level down, a couple of huge trucks are hauling coal. This is the part of the mine where the fires are, says Tom. He describes the mining here as 'fucking challenging, gas and voids to deal with all the time.'

Tom explains how pipes are bringing water from the sump to fuel sprinklers that spray over the heat zones. 'It's kind of like watering a garden, we keep doing it until there's no more smoke.' Across the pit, it looks like smoke is coming out of the ground, but Tom says it's gas – carbon monoxide, hydrogen sulphide, and sulphur dioxide. 'They're fucking lethal. All the guys have to wear gas detectors.'

'If that's gas that I can see, where's the fire?' I ask.

'In the coal seam, about 30 to 40 metres underneath,' says Tom. 'It's just found a way to the top via cracks.'

Because the fires don't have a decent supply of oxygen, the coal is only partially combusting. And the partial combustion process is generating those gases. It's not a normal part of mining. 'It's one of the weirdest coal mines in the world,' says Maz.

The day before, we were seeing flames coming out of the bush, now we're seeing coal on fire. 'World on fire, fuck,' I say.

'You used to be able to see flames here too,' says Tom. 'A long time ago, further down the gully, the fires had broken through the surface, you could look inside this underground chamber, it looked like the bowels of hell, it was this deep, red orangey thing . . .' he trails off.

Tom and his team have scanned old mine plans, but they don't know the exact location of the mine shafts and tunnels.

Some old voids have been eliminated through what he calls collapse blasting, but others remain. To make working safer, all the trucks and diggers have high-precision GPS. 'They've got a screen that tells them where the old underground workings are, so they don't end up falling in a hole.' There are also pockets of ground designated as 'no walk zones', he says. 'If there are any doubts as to what's under the ground, we'll say yes to machinery, because you've got a protective cab around you, but it's too risky to have people on foot.'

This pit is where the high-quality coking coal comes from – the 'good' coal that's exported around the world and blended with other coals for use in steel production.

'It's the best coal in the world,' says Maz.

'Are you proud of it?' I ask.

'I'm proud of the West Coast coal,' she says. 'A lot of coal around the world is like a bulk commodity, but this isn't. To me, it's more akin to gold, or another precious metal.'

'When you think about it,' says Tom, 'for the Indians and the Japanese to send a boat from the northern hemisphere to us, rather than get all their coal locally, tells you something about this coal.'

On the far side of the pit is a small black cliff. 'It's the main coal seam,' says Tom. It's about 10 metres thick and topped by about 30 metres of sandstone caprock. On top of the cliff we can see a pile of rocky rubble. 'We've just stripped that recently. Now another digger will come and clean that rock off, reveal a nice block of coal, and then we can dig it up and send it to the ROM.'

Before we go, I ask about the pink rock around the edges of the pit. It's some of the sandstone caprock that's been superheated by the underground fires. It's like it's gone through a process of metamorphism – something that usually takes millions of years – to change its texture and colour. 'We crush it up and use it for roading,' Tom says.

Tom's got one more place to show us. We drive over a rise and down a steep slope, like we're about to drive into the sea. A red Hitachi digger, a big machine, is working the slope. 'We're moving the dirt off this road to expose a coal seam about 30 metres down,' Tom says.

It looks like a crazy place to be mining coal, but over millions of years tectonic movement has uplifted and folded the coal seam. 'We're chasing the coal down the hill,' says Tom. Mining down from the top of the hill is 'quite painful'. It's an expensive area to work, but the quality of the coal makes it worth the cost.

Geologists have been studying the coal here for generations – gathering information about rocks exposed at the surface, from holes drilled into cliff faces and bedrock, and from deep inside the mines themselves – to create a three-dimensional picture of the mountain, with black seams of coal winding their way through the strata, sometimes exposed at the surface, sometimes tens of metres below, sometimes disappearing abruptly, the sediments deposited millions of years ago uplifted, tilted, folded, offset by more recent earthquake faults. Once they're ready to mine, anything above the coal is considered 'overburden', a problem to solve. Geologists call the rock here 'Brunner coal measures', layers of sandstones, carbonaceous shales and coal seams, mostly deposited during the Eocene, 34 to 56 million years ago. Below the coal measures is an ancient basement rock that includes granites and greywackes. The coal here is young, Tom tells me, compared to the Australian coal, which is hundreds of millions of years old.

This coal was deposited in a cooling world, about the same

time the Antarctic ice sheet was growing to cover the continent. Forty million years ago, a swampy forested landscape sank into the ocean, and was covered by marine sediments – sand, mud, shells – which buried and compressed the organic material beneath it. Like everything this side of the Alpine Fault, it has been uplifted. A 2.1-metre-thick coal seam sighted in the Mōkihinui River in 1862 first alerted prospectors to coal on the West Coast. We have been chasing it ever since, first chipping it off cliffs, then chasing it underground, now digging it out of massive pits, to get every last bit. So we can burn it and release its carbon into the warming atmosphere.

It's windy and cold and we can see across the Tasman Sea to a blue horizon. Below us, Tom points out the Mōkihinui River mouth, the Ngakawau River mouth, Westport to the south. My gloveless hands are frozen and I wish I'd come better prepared for the cold. It's 8°C but the wind chill is taking it closer to zero.

Maz seems unbothered by the cold. 'I'm feeling a bit nostalgic. I'd forgotten how beautiful it is up here.'

Thinking about Antarctica and coal makes me think of global warming, melting ice sheets, rising sea levels, all of which are scarier than stories about lakes of fire and pitchfork-wielding devils.

As we drive across the plateau, up and down slopes, sometimes the barren landscape changes, and there's an area planted, in a neat grid pattern, with subalpine shrubs and grasses. Other places have harakeke, toetoe, tussock, hebes. Tom explains how any materials moved to get at the coal – sandstone capstone, weathered granite bedrock, topsoil – is sorted and stored until an area is ready for rehabilitation. 'We stockpile everything.' When the landscape is rebuilt, it includes a cap of weathered granite that the plants thrive on. There are biosolids in the mix too, that come over from the east coast, he says.

'Biosolids – what's that?'

'Poo from Christchurch,' says Maz.

'Yeah, basically. It supercharges the soil.'

The planting is contracted out, Tom explains, but all of the shaping is done by his team. 'We have a really big focus on ensuring we hit our rehabilitation targets. We try and do 20 hectares a year. It helps lower our bond.' Part of the deal of taking this mine over from the government-owned Solid Energy was putting up a bond to account for any further environmental disturbance. 'It's a bit like a student flat,' says Tom. 'If we leave the place in a worse shape than it was when we took over, we won't get our bond back.'

Maz admires some planting. 'Beautiful, nice bit of rehab there.'

'I designed all this last year.' Tom seems proud of his work. This whole area used to have a 20-metre-thick coal seam under it, he says. Now, only the tidiness of the planting shows where there used to be mine workings. 'In 100 years' time you won't even know there was a mine here.'

We drive up a wide gravel road, past a sign that says *Caution: kiwi in area*. There is low scrub – mānuka, flax, bracken.

'Are there critters up here?' I ask.

'Yeah, we've been releasing the *Powelliphanta augusta* snail from the breeding programme they've got down in Hokitika.' The rare giant land snails have spent more than a decade in captivity, living on beds of moss in refrigerated containers, after most of their habitat was destroyed by open cut mining. Now they're being released back into their old habitat, into newly planted, newly landscaped land. 'They have little transmitters on their shells, to see how much they move,' says Tom.

As we drive on, the carefully remediated, regimented planting gives way to a remnant pocket of bush, flaxes, mānuka, podocarps.

'Is that bit of bush going to stay?'

'Yeah, that stays. That block's dogshit,' says Tom.

'Dogshit?'

'The coal. That's a technical term. The coal's no good.'

The pockets of remnant bush, the remediated plots, all look pretty good. I'm impressed. But I want to know if any environmental groups are objecting to what's going on here.

'At the moment they're leaving us alone,' says Tom. 'They're directing more action at Fonterra and the rail network rather than protesting at the mine gate.'

We drive through an area like a little village, from the days when the miners and their families used to live on the plateau and there was a church, a pub, a school. One of the biggest buildings, Maz tells me, is the bathhouse. This was the miners' first stop after work, where they would remove their dirty clothes, stand side by side beneath the shower heads, and wash the coal dust off their faces and arms before returning home for dinner. It's on Bathhouse Road. The other street names – Soho, Manchester, King's Cross – seem to be named after the red-light areas of various cities.

'We had to move away from that, we got told off,' says Tom. Times have changed in the 18 years he's been here. I ask if there are many women working on the hill. Tom tells me there are women driving trucks, in dispatch, surveying, nursing, admin, accounts.

'Why do they have a nurse up there?'

'It's tough doing shift work,' says Tom. Compared with the physical work of the early miners, most of the mine workers today spend their days sitting in a vehicle. 'It's sedentary, there's lots of eating, overwork, bad sleep.'

When we arrive at Cypress we hit the boundary of the coal Stockton is consented to mine. We can see the edge, a black cliff.

'The coal's just sitting there,' says Tom, 'but we can't touch it.'

Cypress has 'changed heaps' says Maz. 'It was beautiful wetland meadows and now it's a big fucking series of holes.'

Around a prefab building is a ground cover of small coal chips.

'You could take some of this home as a garden feature, Maz, a little coal path,' I say.

'Yes, that would be very nice in St Mary's Bay, Auckland.'

I want a piece of coal as a souvenir. Tom finds a bit of fabric, an old rag, in the truck and wraps up a piece for me. We talk about coal for naughty kids at Christmas, but this lump isn't for my kids. It's for my rock collection. One day, I expect coal to be a novelty, and I hope the story I can tell my grandchildren about this piece of rock will be a good one. I hope we've done the right thing by then.

As we head back, I have another go at talking to Tom about climate change.

'Do you worry about it, think about it?' I ask.

'Not in a huge capacity,' he says. 'I mean, I'm conscious of it. Obviously we need to do what we can do to slow it down. The elephant in the room is population growth really.'

'Oh. The population growth argument, it's not really . . .' But I can't get a word in. Maz and Tom are both talking about how they are aware of climate change and concerned about it, but they don't actively worry about it the way I do.

'I wonder if it feels different if you don't have kids?' says Maz.

I want to tell them the population growth argument is bogus. The planet couldn't cope with even half our current population if everyone consumed as many resources per capita as we do in New Zealand, Australia, the United States, but there are other ways of living, sustainable ways of living. There will be room for more people if we stop farming animals, stop driving private cars, stop producing and consuming so much *stuff*.

But it sounds like Tom's on track, even if we're starting from different points. 'Obviously the human race has to move into a different way of operating, full stop,' he says. 'The way forward is public transport, high-density housing, high-density agriculture – we've got to feed ourselves. But there's no bloody free lunch, you're going to leave a carbon footprint of some kind. All the technology we use, that's got to come from somewhere. And that includes minerals from the earth – they don't grow on trees.'

'I know you're mining coking coal, but what do you think about mining coal just for burning, for heat, or electricity?' I ask.

'Well, if they can find a way of producing heat without using coal then great, I'm all for it. But the solution is not one thing, it's going to be a spectrum of alternatives to thermal coal.' Tom's been listening to an online lecture series on energy which points out that each country has to come up with its own way of solving heating and electricity needs. 'Solar panels are great on a small scale, but New Zealand just doesn't have enough sunlight hours. Same with wind. You can't put wind turbines everywhere because it's not consistent enough.'

In New Zealand, though, I point out, we're already doing

pretty well, with more than 80 per cent of our electricity coming from renewable sources.

'We're lucky to have so much hydro,' says Maz, 'rather than being reliant on coal like Australia'.

But most coal alternatives – whether a nuclear power plant, a wind farm, a solar panel array – require concrete in their construction. 'People don't realise you use an awful lot of coal to make the cement that goes into concrete,' Tom says. He read something recently about a new method for cement making that replaces some of the coal with potash.

'Chernobyl's kicking off again, have you read about that in the news?' says Maz. 'It's heating up and they can't get to it because the whole thing is buried.'

We've all watched the recent TV series *Chernobyl*. 'Do you know what, it was the nuclear version of Pike River,' says Maz. 'The bullshitting that went on and the trying to look good in front of all the bosses, all of that. Read Rebecca Macfie's book, it's stunning. One of the big lessons that came out of the Pike River Royal Commission was about mining education, making sure that anyone who works in the extractive industries has a verified level of competence.' She's now on the Mining Board of Examiners and says she's seen standards lift across the whole extractive industry.

What next for the mine, and for Tom? He's talking about doing his quarrying ticket. Talking about the infrastructure projects to come. Post-COVID, the price of coal and aggregate is going to go through the roof, he reckons.

'Is there an end in sight? A time when the mining up here will be finished?' I ask.

He says it all depends on what happens at Denniston, where there's another pocket of coal they want to mine. 'If we don't get the Denniston Road through from Cypress, then things are looking pretty hairy for us around 2027.' Once the mine closes, he says, 'The land will be rehabilitated, planted and handed back to the community.'

Back at the carpark, we return our hi-vis and helmets to Merv.

'It's not his real name,' says Tom as we say goodbye. 'But he is like a Merv. Not a bad geezer, actually. He's all right. Very talkative.'

Before Maz and I get back in our car, I go to investigate a sign I can see partially hidden by bush at the edge of the carpark. *Mary McGregor Memorial.* There's an information board:

> Welcome to Stockton, once a thriving community established earlier this century by pioneers, mostly from Scotland, Wales, Ireland and England. Many of these people migrated to New Zealand in search of work at the end of the First World War. These early settlers formed the Stockton community and worked long hard hours in the underground mines further up the plateau.

At its peak, there were 500 people living here. But the big strike during the 1950s 'was the beginning of the end. As the work dried up the miners moved away in droves.'

There's a photo of Mary McGregor. She emigrated to New Zealand from Scotland with her husband and ten children in 1923. They lived in a rough-sawn two-room bach. 'Six weeks of rain and mist welcomed them. Jobs they had been promised did not eventuate.' What's interesting to me, though, is that Mary was a midwife, like my great-great-grandmother Marian.

> Mary had been a midwife in Scotland and seeing a need in Stockton for a midwife and a nurse, began regular rounds

with Dr J Simpson who was the coal company's doctor. They worked together for many years. As there were few roads and even less vehicles, she walked many miles across rugged terrain and corduroy paths to tend people in Mangatina, Mine Creek, Millerton, Middle Break, Hector, Ngakauwau and Granity. On occasions she went on a bus to Birchfield and up the Buller Gorge.

Mary McGregor died at 74, which is what I think we'd call a good innings for someone of that generation. My great-great-grandmother Marian, who also lived to 74, was from an earlier generation, arriving in New Zealand in 1864. Her midwifery was of another time, and I know nothing about it.

As we drive down the road, I'm thinking of how different our lives are today, and the costs of our so-called progress.

I learned about climate change at university in the late 1980s and early 1990s. I've been working with climate scientists, and writing about climate change, for two decades. Coal is the single biggest contributor to climate change, responsible for nearly half the extra carbon in the atmosphere from burning fossil fuels. Greenhouse gas emissions also come from the petrol we burn in our cars, the gas we use for heating and cooking, from burping livestock, and sometimes from methane gas that seeps out of the ground and into the atmosphere. The latest science says we have mere decades to cut net carbon emissions to zero if we want to limit global warming to 2°C by 2100, and avoid the most catastrophic climate change, ice melt, sea level rise scenarios. Based on current commitments, we're heading for a 2.7°C increase above pre-industrial levels by the end of the century, but the more we can bring that down, the better.

Tom and Maz's matter-of-fact statements about the coal, and why it's okay to keep mining it, make sense; it's kind of logical if you think of it as mining coking coal to make steel for wind turbines. And it's interesting. I love big machines and it's like something from a Richard Scarry book seeing the production chain from excavation to blending to hauling by truck, ropeway, train, ship.

But it's all details. The bigger story is we need to decarbonise, find ways to live that don't depend on fossil fuels. In the car with Tom and Maz I had thought about delivering a lecture on all of this from the back seat, to try and impress upon them the urgency of the situation, but figured they'd heard it all before, know what needs to be done, and are on board with most of it. Responding fairly to climate change, though, will mean that those of us with plenty might need to learn to live with less, and sometimes that's the hardest sell.

5

Apocalypse

We'd been friends since pre-school and best friends since the start of high school, but before we became Christians, before our hearts and eyes were opened to Satan's grip on the world, Maz and I were angry, because everything was fucked.

We were angry about the System. We were angry about fascism and racism, but mostly we were angry about sexism because we were told 'Girls can do anything!' but that wasn't true. 'Anything' didn't include ditching our girls' tartan tunics for regulation boys' grey shorts; the male chauvinist pig teachers sent us home to change. 'Anything' didn't include getting the Most Improved Player cup in soccer; that was for a player in the boys' team. To vent our angry feelings we went to women's self-defence classes, where we learned how to fight using our keys, elbows, feet and fists and were given stickers that said *All men are rapists* and *Women who protect themselves against the violence of men are acting in defence of all women.*

Sometimes we harnessed our anger to take a stand on things. We protested against the 1981 tour by the racist white rugby team, the Springboks, which made people in our rugby-loving school hate us. 'Support the tour, you fucken bitches,' said Robyn when we walked past her wearing black armbands to signal support for Nelson Mandela and the anti-tour movement.

Our female relatives, women a generation or two older than us, were angry too. They led us on protest marches, and wrote manifestos, and shouted. Some of them were arrested, and one formed a protest movement called Women Against Rugby. We also marched in protest at American nuclear ship visits because we worried a lot about nuclear war, even though our homes were Nuclear Free Zones.

But now, as Christians, we were happy. The world made more sense this way.

But we still didn't fit in. If our new Christian friends thought we were *different* to the other girls, it was at least partly because of the way we looked. There's a photo of us from 1983, just before we became Christians. I'm wearing an oversized black fisherman's jersey over a white T-shirt, the strap of an Army Surplus canvas haversack across one shoulder. My head is turned slightly towards Maz, chin down, eyes up, and I'm scowling. Maz is leaning in towards me, a denim jacket – sleeves cut off – worn over a hand-knitted purple jumper. Her head is similarly angled but instead of a scowl she's wearing a cheeky, impish grin. She looks cute. I look a bit scary. The occasion for the photo was our new hairdos, ratty buzzcuts executed by electric hair clippers, with a lack of precision, and much hilarity. The sun makes golden highlights in our medium-brown hair, about half a centimetre long all over, except for the 'fringe'. The longer front strands of Maz's hair are woven into three tiny plaits, one hanging down in front of her right eye in a way that would infuriate our teachers when it grew long enough for her to chew the end in class. Over my forehead I have a thin fringe of hair that's curling to one side. I hadn't realised until my hair was this short that I had a cowlick.

The hairdos, as well as entertaining us – the foundation of our friendship was to make each other laugh – were a 'fuck you' to our teachers. At the start of the school holidays we had spent all our savings at Peter Zidich hair salon in Wellington, having our hair bleached white then dyed Ultraviolet and Bahama Blue, new plastic-bright colours that had just arrived in New Zealand.

When we returned to school looking awesome, one of the head teachers – eyebrows plucked into a thin arch and hair dyed tart red – called us into her office and told us we were not welcome in school with blue and purple hair; the teachers who had seen us arrive were refusing to have us in class. We went home and our mothers came into the school to defend us. 'They're not taking drugs, they're not having sex with boys, they're not failing their classes. Why are their hairstyles so important to you?'

We obliged by having the colour stripped, returning to school with platinum blonde locks. But we liked messing with our hair, and one afternoon we took to it with electric clippers, emerging from the bathroom, our blonde hair replaced by short brown fuzz with patches of scalp showing where we cut too close. In the kitchen, our mothers were drinking wine together.

'Jesus Christ,' they said before bursting into laughter.

When we told them not to take the Lord's name in vain, because we'd already started hanging out with Christians, they laughed harder.

We returned to school the next day with smug smiles, compliant with the rules but with hairstyles that were even more extreme than the blue-and-purple salon jobs. We liked that our mothers had defended us.

Mostly, though, we felt disconnected from our mothers. They were busy saving the world, talking about women's lib, teaching underprivileged or handicapped kids, and dealing with our younger siblings. They trusted us, but it felt like their attention was usually elsewhere.

Our school in Petone was what was then called a 'melting pot'. There were students whose first languages were Greek, Hindi, Korean. There were refugees from Kampuchea. There were students from every Pacific nation. The deputy headmaster, who was Māori, made it very clear to us that the Māori students had mana and were to be respected. Many of them were from important families, had last names that matched the names of the streets around us, were descended from the tīpuna who had founded the nearby Waiwhetu Marae.

Our own families had been in Aotearoa long enough to have lost touch with European relatives, but we weren't Indigenous. With no religion and no sense of our culture – this was before people like Michael King were writing about what it meant to be Pākehā – we struggled to know who we were, other than that we were Pākehā, even if Maz did take 'Māori' as one of her options (I took French), even if Maz sang with what was then called the Poly Club, and even if my classmates told me I had 'Māori eyes' and quizzed me about my whakapapa. So we found our culture in music. Music gave us a way to define who we were, delineate our allegiances, declare our enemies.

At our school, you were either a Punk, a Rasta, a Trendy or a Normal. We were unbothered by the Normals but could not abide the Trendies, with their bland conformity, their expensive camel Nomads and black bomber jackets. Most of our brown classmates were Rastas, and when a classmate died, hit by a car as she was crossing a road late at night, they assured us, 'She's with Jah now.' We had respect for the Rastas.

Maz and I, of course, were Punks. We had always been a bit weird. When we started high school we wore Roman sandals and ate Vogel's bread sandwiches and had no money for the school café. But now, with our home-cut 'Chelsea' hairdos, we really stood out. Perhaps the Trendies and Normals found our hairstyles alarming, but outside of school our haircuts gave us credibility. We knew about punks from *Rip It Up* and *NME* and we made it our mission to find some. In the weekends we would catch the train into town and then scamper around the city, or drive around our neighbourhoods in Maz's beat-up orange Morris Oxford van, looking for punks and waiting for stuff to happen. My favourite outfit was a floral crepe de chine skirt, roughly cut and hemmed from a 1940s dress, an obscure band T-shirt, my unravelling and oversized fisherman's jersey, black laddered tights, big boots, and a near permanent scowl that hid my shyness. Though we rejected 'girliness' – our only other nod to feminine norms was shaving our legs – being skinny was important, and there was an endless stream of diets. I tried the bread diet, the Israeli Army diet, the Atkins diet.

We went to gigs to listen to bands with names like Flesh D-Vice and Skank Attack, and at home we burned incense and read books by Carlos Castañeda and wondered who we had been in past lives. We read our parents' copies of *Down Under the Plum Trees*, which made us wide-eyed and blushing, and the *Whole Earth Catalog*, which inspired us to experiment with ground nutmeg and dried banana pith. Sometimes, though, we just lay in bed, listened to Joy Division and the Cure, and cried.

When Maz started going around with Theo, we went to his

house at lunchtimes to eat handfuls of broken biscuits out of the box from the Griffins biscuit factory and then Maz would kiss him with biscuit crumbs in her braces while I talked awkwardly to his friend. Sometimes we'd smoke a joint. Maz didn't much like it, she said it made her feel paranoid, but I liked the way it made everything woozy and mellow. We started drinking that summer, at a family garden party, when I took two bottles of Marque Vue into the tent we'd pitched beside the jasmine bush and made a concerted effort to get drunk. It worked, but I was a miserable drunk. After the giggling stage, then the vomiting stage, I sat in my bedroom with my friends and scratched long jagged lines down my right arm with a large safety pin, the cool sharp focus of the physical pain pulling me out of my emotional pain. I'm sure there were deeper issues, but I was 15 and I liked a boy. I didn't know if he liked me. It made me mental.

The next day, after the hangover had lifted, we both agreed that getting drunk was much worse than smoking pot.

After months of hanging around the margins of the Wellington punk scene, trying to find Our People, at last we met some other schoolgirl punks. They looked like the punks we'd read about in magazines. One even had a mohawk and tiny plaits, like Annabella from Bow Wow Wow. On Friday nights we would catch the train into town then follow them up dark alleyways between rows of hillside houses, past Gothic churches and up decrepit flights of concrete steps to a place they called Flagstaff, which we could never find again on our own. At Flagstaff, a pile of punks, real punks, lay around on the grass, a cassette player smashing out the Dead Kennedys. They drank from bottles of cider and passed around joints. A couple of young guys with shaved heads bounced around the edges, play fighting, throwing things, and laughing as they fell on people.

After they got used to us hanging around, some of the older punks started telling us where the parties were, and we went to a house they called '301', where it was dark and loud, and some of the boys stopped being fun and started leering at us. We decided we didn't like parties. But we liked going to see bands play. And we liked getting out of it.

I now realise I had crippling social anxiety. I was shy and self-conscious and often found myself unable to speak, even if someone spoke to me. Alcohol helped. Not too much, but a swig out of Mum's brandy bottle before I went out. And Maz helped. I was happiest when she was half a step ahead of me, navigating our social interactions, making things easier.

One time, at Suburban Stomp, where some bands we liked were playing, for no good reason that I can remember, I drank cheap white wine then took some prescription pills that someone had told me could make you high. The combination of wine and drugs made the room spin, and I got to the bathroom in time to vomit and vomit until I felt so weak I couldn't get up. I lay underneath the basin, curled in the foetal position, on a wet floor scattered with paper towels, feeling worse than I had ever felt in my life but also aware that I was having an Experience.

'Did you take some of your mum's flu medicine?' said one of the mean girl punks while she stood over me to wash her hands. As she spoke, she pushed her boot firmly into my head.

'Yep,' I said quietly.

Then I realised I only felt nauseous when I breathed in.

'She's gone blue!'

'Is she dead?'

There were boys' voices now too.

'Okay, she's alive. Call me if she dies again,' said our friend Peter from Destructive Adolescents, and then left to play another set. We made it home that night. We didn't talk about it at the time, but later we both agreed that was the night things started getting a bit out of hand. We decided to rein things in a bit, maybe stay away from alcohol, maybe spend more time with the increasing numbers of our friends who had stopped scowling and stomping and were now walking around with happy and peaceful looks on their faces.

6

Bedrock

'You know Granity is falling into the sea, yeah?' says Maz.

I look past the row of houses that separates the highway from the stony beach, and nod. As sea level rises, some coastal towns are making plans to retreat, to move buildings, walkways and roads inland, and let the sea do its thing to the coastline. In Granity, there's nowhere to retreat to. It's home to about 160 people living on a tiny strip of land between the steep-sided hills and the encroaching sea. As you head inland from the highway, there's the row of houses, the railway line, then the hills. We drive past St Peter's Anglican church – *I am the way the truth and the life: Jesus* says a sign – and the Granity Judo Club and then a startling display on an unhitched trailer. A life-sized naked female figure, with large round breasts and no hair, reaches her arms high above her head to hold – or throw? – an angry-looking baby. Orange flames reach as high as her raised elbows. Beneath the flames are the words *GLOBAL WARMING*. The placement of figure and flames, and the fierce look on the woman's face, give the impression she is about to hurl the baby into the flames, but I doubt that's what's intended.

We're heading for Karamea to stay with an old school friend.

I haven't seen Milly since high school, but Maz connected with him when she lived on the Coast and now he's texting – there's talk of a drive up the road to see the sunset, maybe a walk to the Ōpārara Arches, whitebait fritters for breakfast tomorrow.

At the north end of Granity, we park on a gravelly verge and walk over a mound of earth. There's a pīwakawaka flitting about and gulls circling over the grey-green water. I walk south along the steep beach, which is covered in rounded pebbles of speckled granites and diorites. The sections backing onto the beach all have barriers. There are fences made from driftwood, others of traditional wood and wire mesh, one with fancy gabion baskets, and one concrete block. Further along the beach, a stretch of houses is protected by a long pile of sandy coloured boulders, a rock revenant or riprap. There's a gentle roar of the ocean, a constant reminder that the tide is coming in. One section, lower than the others, has no fence. The sea has washed a river of stones and huge bleached logs over what used to be the lawn and up around the house.

I walk until I reach a tumble of broken concrete and mangled reinforcing wire. It's impossible to tell what it used to be. A sea wall? A pier? Remnants of an earlier civilisation; the concrete and iron age. I take some photographs then turn back towards Maz, who's sitting in the distance, hood up, looking down at her phone. A black-backed gull flies low over the waves and a yellow glow of morning sun starts to appear through grey clouds. As I'm looking at the sky a wave surges up the beach towards me. I run but the foaming water seeps through my boots to wet my socks. I shout out – 'Maz!' – but she can't hear me above the tumble of rocks being dragged back by the undertow.

I join her, laughing now, and we look along the beach, to where a digger and three people in orange hi-vis are

75

working. There are works planned here – a news article says the government has promised funding for a seawall between Ngakawau and the north end of Granity. Locals, though – one of those interviewed describes the town as being populated with 'rednecks and creatives' – are asking about the rest of Granity. The seawall will protect only five houses, three of which already have rock barriers.

An oystercatcher pecks at the sand, and I think of food. It's 11:43am – high tide isn't for another three hours.

We drive south to Waimangaroa, where we saw a sign earlier – *Fresh homemade pies cooked daily*. It's an outdoor café, with a food cart and a few picnic tables next to some sort of post-apocalyptic sculpture garden. A man is talking to the woman in the food cart. 'The worst you can do about it is have kids,' he's saying. I wonder if he's talking about climate change, and arguing the population line again, and listen as he continues. 'Cos they're little assholes all the way through. Then, just when they're useful, they grow up and leave home. You're better off getting a dog.' I laugh.

It's been a busy summer, the woman in the food cart tells us when we ask how business is going. We're aware things have been tough for the Coast, which relies on tourist dollars, and are happy to be spending our city folk salaries here. We order tea and pies – chickpea and pumpkin for me, steak and cheese for Maz – then go for a wander. The sculpture garden contains derelict machinery, wooden carvings and vegetation. A man called Woody makes the art, we're told. We pass a rusted-out truck, a tractor, larger-than-life human figures carved with tā moko on faces and buttocks, driftwood carvings behind glass screens. Under a corrugated iron roof is what looks like an ancient waka. Beneath and between the sculptures the garden is wild with ponga, nīkau, grasses and pond life. We come back smiling, buzzing. We find a place to sit, and nod a greeting to the man, who's now sitting at the next table.

'Where're you ladies from?' he asks. Auckland and Wellington, we say, then Maz tells him she used to live in Westport and they have the conversation in which they identify people they know in common. Things have changed on the Coast since Maz left, he says. House prices have doubled in the last few years. 'People are asking crazy prices for West Coast houses, and they're selling.'

We talk about COVID-19, the vaccine rollout, and the new variants we're starting to hear about. 'I probably won't take the vaccine for a while,' the man says. 'I'm not really into it.'

The woman from the food cart delivers our tea, a big pot of English breakfast with the kind of floral bone-china cups and saucers my Nanna used to have. She tells us her daughter is vaccinated 'homoeopathically', and neither she nor her daughter are going to 'take' the COVID-19 vaccine. She knows someone whose kid nearly died after a vaccine – he had an allergic reaction. It's put her off, she says.

I listen, and nod, but when the man starts to query whether the Pfizer vaccine is really a vaccine, 'since it changes your cells', my respectful and curious listening reaches its limits. 'I'm going to do a little plug,' I say. 'It *is* a vaccine. And it had been tested on millions, if not billions, of people by the time it got to New Zealand.'

He agrees there's a lot of misinformation out there. 'We've had the leaflet drops. They say it's going to pre-programme you to die,' he says. 'Will you guys take it?'

'Absolutely,' we both say.

'Are you guys nurses?' asks the woman.

'No, but we're both science-trained,' I say.

'Personal choice,' she says with a shrug, and walks back

to the caravan to check on our pies. None of us are eligible for the vaccine yet anyway – we're not MIQ or healthcare workers, or in the vulnerable populations or age groups – but our turn will come later in the year.

Talk turns to conspiracy theories and the way they intersect and tap into human fears and anxieties. The thing I worry about is climate change, I say. And I worry more about COVID-19 than the vaccine. I talk about my Te Pūnaha Matatini colleagues in Auckland and Christchurch, the ones doing the COVID-19 modelling, and how much I respect and trust them and listen to what they have to say.

The man takes my point. 'Anyone who left school at 15 and has access to a computer can say something and think they're an expert,' he says. 'What do you guys do?'

'I work at a university,' I say. It's too hard to explain any more than that. I'm not teaching this year and people generally think that 'university lecturer' means teacher rather than researcher, writer, student supervisor, and everything else the job involves. Besides, I spent the last three years managing staff, and I'm still in recovery from that. I'm not quite sure what I 'do' anymore. Maz says she's an engineer.

'I kicked off Bathurst here when they first came to town,' she says, 'before they took over Solid Energy. Got the consent for the mine in Denniston. Now I'm in Auckland and working in construction up there. Bit of a change.'

The man tells us he's between jobs at the moment but studied philosophy and earth sciences at university. Another thing we have in common.

A dog turns up and I call it over to give it pats. The man stands to leave and nods goodbye to us. 'Nice to talk to some people who are intelligent,' he says as he walks off in his gumboots.

We're silent as we sit eating our hot pies. A train rumbles past. Dirty carriages, dozens of them, black with coal dust.

We're due in Karamea, but first I'm keen to visit the Denniston Plateau, which is a 15-minute drive up the hill from Waimangaroa. Like the adjacent Stockton Plateau, the Denniston Plateau is a bleak and rugged landscape with a coal-mining history that dates back more than 100 years. It is also the site of the Denniston Incline, the precipitously plunging ropeway made famous in my aunt Jenny Pattrick's bestseller *The Denniston Rose*. Maz and I first came up here together in the winter of 2011 when, in high spirits, we disobeyed the *Warning! Do not attempt to walk the incline* signage and scrambled down the vertiginous slope, bracing our knees, climbing with hands and feet, grabbing trackside saplings or each other to stop ourselves slipping, loose gravel and chips of coal bouncing down the slope before us. When the incline became vertical, and broken tracks crossed a deep forested ravine, we took heed of the *Danger! Unstable structure* sign but, rather than climb back up the steep slope, pushed past the one that said *Track closed* to traverse the slope in the direction of the road down the hill. About a kilometre along, we found a huge, fresh landslide, and held tight to a guide rope that kept us from joining the dirt and scree at the bottom of the slope.

I had visited the Coast because I was feeling unhinged and had reached out to Maz for life support. Just out of a nine-year part-time PhD, during which I'd birthed and cared for three children, I was exhausted, and had started acting out of character, smoking cigarettes, going to gigs – backslidden from my domestic life in search of an identity other than mother, wife, student. Maz helped me to laugh, to ground myself, to stop overanalysing things and just get on with it. On our Cyclone Ita trip, a few years later, my husband Jonathan and I took the kids up to the plateau where the twins played a game they called 'apocalypse', hiding and sheltering amongst broken brick walls and mist-shrouded, orange rusted machinery.

Today, we walk over the broken concrete pad of the Brakehead, amongst the abandoned machinery and structures, then I clamber into a restored coal wagon and think about the terrifying ordeal of travelling up and down the incline in one of these wagons – between 1880 and 1902 it was the only way up to the plateau – and empathise with Rose from Denniston, who, once she came up, refused to ever travel down.

It's getting late, and we have more driving to do. On the 1pm news, we hear that on the east coast the town of Hinds is surrounded by water, Pines Beach is being evacuated, the State of Emergency holds. Everyone living in the Coopers Creek catchment is told to evacuate to higher ground.

After the news I put the playlist on and we sing along to 'Anarchy in the UK'. There's coal smoke in the air. I watch the grey-green sea, high tide approaching. After a few minutes on the coastal highway we're in Ngakawau. The aerial ropeway is bringing green bins of coal down the bush clad hills from Stockton to a complex of industrial buildings, chutes and conveyors. Black pyramids of coal sit waiting for the train. Much of it looks like it was built in or since the 1950s, when the aerial ropeway was installed, but there's also a long high-ceilinged concrete building with huge arched windows. *Est 1905* it says on the side in raised letters. This is the end of the Stockton complex operation, explains Maz. From here, the coal gets loaded into trains and hauled down the coast to Stillwater Junction, just inland from Greymouth, and then over the hills, up inclines and through tunnels, to the east coast port of Lyttelton. From there, most of it is exported for steel making. It's ironic that the only trains still running around here are the coal trains, I think, when trains are such a sustainable means of transport. I grumble about the loss of passenger trains as we drive north, over the Karamea Bluff to Corbyvale, past huge white painted letters on a mossy road cutting – *MOISTY*. We laugh, and Maz asks me to take a photo. The sky is dark and high, our side of the weather system dumping rain on the east coast. There is snow on the mountains ahead.

Our mixtape has reached Bob Dylan's 'Gotta Serve Somebody' and we start talking about our Christian days. The way we would copy the more established Christians' body language during worship. I favoured the hands gently open, at waist height, in front of me. Maz was more dramatic – arms up in the air, head back, her voice soaring.

As we pass a wetland, a sign warns us to slow for bittern. And soon we reach a rainbow-striped sign: *Welcome to paradise. Population 575*, though the number has been painted over *650*. We're in Karamea. Driving into town we realise we don't have Milly's address so Maz slows down while I scroll through her phone to find his number. But we don't need to call him. There he is, standing next to a black 4WD at Karamea's only gas station. Maz pulls in, shrieks a greeting to him through her window, then gets out of the car and bounds over for a hug. I follow behind. I haven't seen him for decades. I smile, say hello, wonder about putting out my hand for a shake but he goes in for the hug. It's fine. There's something about people you knew as a teenager that breaks down the barriers. He's tall, lanky, wearing black canvas workpants, boots, a fleece. He has the dirty fingernails of a mechanic and the angelic blond curls of his teenage years have turned into a wild grey mop that he's topped with a black beanie. He has a full beard, completing the West Coast wild-man look. I realise I'm dressed kind of the same as him though my hair is neater, and I may be wearing lip gloss.

We follow Milly out of the gas station, down a couple of roads, and up a gravel driveway to a wooden house with solar panels on the roof, a large sleepout, and a grassy section on which are parked a van and several utes. Under a large lean-to roof, shelter to bicycles, plants on shelves, and a table laden with pumpkins, we climb wooden steps to the house. Inside, Milly shows us to a room with a metal tube bunkbed and bare mattresses. I throw my sleeping bag onto the top bunk, familiar with adult bunk sleeping from geology field trips and Scott Base visits, though the last place I shared a bunk with Maz was probably a 1980s Christian camp.

Milly is an exceptional host and has stocked up on packets of chips and two bottles of red wine – both for us, as he no longer drinks. We exclaim how generous he's being, but he won't have it, thanks us right back. 'The gift is in the giving,' he says. We open a bottle and stretch out on the blanket-covered couches to chat and consult our phones while Milly busies himself peeling vegetables in the open-plan kitchen. I talk a little bit about my writing project and Milly puts down his peeler and pops into the next room. He comes back with a couple of books, David Suzuki's *The Sacred Balance* and Rumi's *Bridge to the Soul*. I open the Rumi and read a poem, 'The dance of your hidden life', while Milly and Maz continue to chat. Something about the way Milly talks and moves – slow and considered – is very calming and what with the poetry, the couch, the fireplace, the wine, I can feel myself falling into a state of deep relaxation. One I haven't felt in a while.

I realise that the years of parenting, working, fighting the patriarchy, the relentlessness of every single day, have worn me down. They say that if you're a woman, once you get to a certain age you're either invisible or you're the CEO. Maz and I have both worked in male-dominated fields, been the boss, smashed through those glass ceilings then put a hand down to try and haul other women up after us. But we've also encountered what Helen Clark referred to as the 'wall of men'. When you realise you're the only woman at the table, or when you walk into your first management meeting and are pleased to see so many women in the room, then realise the women are all in administrative positions and all the managers – all except you – are white men. You point it out to them, and they laugh, and agree, yes, it is a problem. When you raise issues of gender disparity and workplace harassment, again you are greeted with nods and concern, and maybe even new policies – but then nothing changes. You never really felt different to the men, never had much of a sense of gender, but you know that most of the men you work with gender you every day. When they apologise for swearing in front of you, when they meet your husband and ask how he copes with the cooking and the kids while you're away on a work trip, when they hit on you. All the years of mansplaining, sidelining, condescension gets really fucken tiring. And then you realise there's a younger generation who tend to put you in a box, because you're what they refer to as a 'middle-aged white feminist' in a position of authority, assuming you must be some kind of agent of the patriarchy to achieve what you have, and you've never felt so unseen.

～

When a teenager – Milly's partner's child – arrives on a bicycle, and asks, 'When are you leaving? Can you take me with you?'

we all laugh. But I get the feeling they're only half joking. This place literally is the end of the road. What brings some people here makes it feel stifling for others.

It's close to sunset, so we grab the open bottle of Shiraz, get into a black Toyota Hilux and Milly drives us north as the sun sets over the Tasman Sea.

'Hurry, we're going to miss it!' says Maz.

'What do you mean, we're going to miss it – we're experiencing it right now,' I say, looking out my window and across the ocean to the setting sun.

'Exactly,' says Milly.

We park up at the end of the road, at the Kōhaihai River mouth, put on our coats, and carry our wine glasses over golden sand dunes to sit facing the roaring waves and watch the horizon, where the sun glows yolky yellow beyond dark seas and below pink and apricot clouds. Once the sun has set, we turn our focus to the waves crashing on the steep sandy beach, the birds, the bush. South of us, the beach stretches down to Karamea. Just to the north is a dark rocky point. It's the end of the beach as well as the end of the road. It's at once dynamic and peaceful. Across the road is the start of the Heaphy Track. Maz and I both admit we've never walked the Heaphy and make plans to do it.

A weka calls from across the estuary.

'He's saying, "This is my fucken territory,"' says Milly.

We talk about how special this place is.

'There's no bad vibrations here. You get to the end of the road and there's nothing there,' says Milly. 'I've seen so many people release their negative energy up here, at the top of the road.'

I ask if he thinks there's something metaphorical about it, about coming to the end of the road. 'I think that's part of it,' he says. 'You've arrived. You've nowhere else to go.' But then he says it's also because we're on granite. My ears prick up. Granite is my favourite rock. It's common in many parts of the world, a typical continental rock, but it's not so common in Aotearoa. We have two granite deposits. Erosion of the 110-million-year-old Separation Point granite, rich with pink, yellow, white and black crystals and exposed at the northern end of Abel Tasman National Park, weathers to give the park its beautiful golden sand beaches. The 375-million-year-old Karamea granite mostly sits beneath the mountains of northwest Nelson, bedrock to the sediments that layer above the granite. Both granites are ancient rocks, remnants of Gondwanaland, our country's genesis as part of a supercontinent. 'It's so old, so fundamental to our geology. It connects us to other continents, Aotearoa's geological parents,' I say. But what is Milly talking about? Why is the granite special?

'Granite is completely neutral,' he says. 'There's nothing to influence it, no magnetic field. In a spiritual sense, you're completely unmolested by negative energy.' He starts talking about the 'Granite Pathway' and I don't really understand what he's saying but I kind of like it. Maz joins in, describing granite as comfortable, energised, like a mother's hug. She loves the energy of it. We get talking about the power of rocks. At some point I remind them that I have a geology degree, though now I can't remember if I studied geology because I was really into crystals as a teenager or if the crystal collection followed the geology degree. Maybe it doesn't matter. I love rocks.

We're talking about how peaceful we feel, and what a contrast it is to where we were, and how we were feeling, just 48 hours ago.

'Life is a progression of moments and if you can get a couple of good ones in every day you're doing okay,' says Milly.

By now, with the wine, the sunset, the company, I'm open and receptive and start writing down Milly's utterances like he's my guru.

Even though Milly is a lot more settled than he's been in the past, it's been a rocky few years for him, breaking up with his previous partner. But he's okay. 'I think I've got a few more rounds left,' he says. Next time he moves, he'll find somewhere closer to the bush and sea. He loves Karamea – 'It's about as close to being in the bush I can be and still get paid,' he says. Even so, there are too many people here for his liking. 'Once my kids can stand on their own two feet, I'll go further in – Scott's Beach? That's what I'll do when I retire. Sit on the beach and fish and drink whisky.' But he pauses and seems to think a bit. 'Who's going to carry me home? Or slide me back into my tent?'

I hold back tears on the beach. Just for a moment I get a sense, a glimmer, of peace that I want to find a way to hold on to.

'You're grounding yourself,' says Milly.

We drive back to Karamea, the bottle of Shiraz between my knees. We can see lights at the top of a hill. 'That's Stockton,' says Maz. 'They're still working up there.'

—

Back at the house, Maz and I eat salt & vinegar chips while Milly cooks dinner. He has a tray of kūmara, pumpkin and potato roasting in the oven and now he's preparing the fish. While he fills bowls with flour, eggs – 'home-made eggs' he boasts – and breadcrumbs, he and Maz talk about the gang, people we went to school with. There's Grant, Spider, Blen. Some of the names are familiar to me but I can't really remember who's who. Our main friends were the art nerds we met in our first year of high school. Our art classmates were funny and nonconformist, and some of them stayed our friends for years, joining us on various missions and escapades. An Easter tramp to Butterfly Creek where Maz forgot the tent, and heavy rain drove us to walk home through the bush at 3am. A boredom-induced driftwood bonfire by the lighthouse at Pencarrow Head, after which Maz and I rode home in the police car that came to investigate, leaving the boys to walk the seven-kilometre gravel road home in the dark. An end-of-exams study-note bonfire on Days Bay wharf, where the wind caught the flaming pile of notes and scattered them into little fiery boats riding across the wave tops, while we screamed and ran from the wind-borne pyres.

Tonight the wood burner, fed mostly with driftwood Milly picks up on the beach, is pumping out the heat. RNZ National is on the stereo. Half a metre of rain has hit Canterbury so far, say the reports. 'Holy shit, we haven't had half a metre of rain all year, combined,' says Milly.

Karyn Hay is doing a plug for a show that comes on later in the evening. 'Remember Karyn Hay?' says Milly. Of course we do. When we were teenagers she was the host of *Radio With Pictures*, which was one of the key ways we found out about new music – punk, new wave, new romantics, local bands, all of which were labelled 'alternative'.

We compare notes about our teenage years. Milly came from a larger family, Catholic. I tell him about my blended family – my mother, sister, stepfather and step siblings.

'That blended family bullshit is really hard . . . and I'm a real placid bastard,' he acknowledges. His partner Steph, who is driving back from Christchurch today, has two kids. His two kids and their mother live in another house in Karamea.

While he's getting out the cutlery to set the table, he tells us about his new obsession – knife-making. He's been collecting wood for years, for the knife handles, and now he's collecting steel. 'I'm building a power hammer,' he says. 'I've been making knives for twenty odd years, but not properly. We used to mine sand off the beach and fuck around with it and see if we could smelt it, but they were just trinket knives.' Yesterday he started making a plinth for his anvil. 'Once I get the powerhouse fixed up and start smacking hot steel around, it will be a different fucken story.'

For dinner there are green beans to go with the roast vegetables and pan-fried fish. I'm impressed with the beans and ask what the gardening is like up here. 'Green beans, garlic and marijuana grow exceptionally well,' he says.

I've bought some offerings for dessert so after dinner we eat Tip Top hokey pokey ice cream out of mugs then nibble at blocks of Whittaker's. I've brought West Coast Buttermilk Caramelised White Chocolate and Wellington Roasted Supreme Coffee. They're delicious.

I ask Milly about the trouble he got into when he was younger.

'Fuck, you name it, I did it. I was on PD for ages,' he says. Offensive language, auto theft, drunk driving. He doesn't do these things anymore, hasn't been in trouble for a long time, he says, though he's had a few run-ins with the local cop, when he went out fishing during the big COVID lockdown.

'I requested a list of my convictions a while back. It was nine full pages. I didn't know I'd done half of that shit.' There's one that's stuck in his mind, though. 'We did beat up that rapist. I did four months PD for that.' The guy they beat up had gang associations. He and his friends had to be very careful about what they did and where they went for a while.

'It was a bit of a spur-of-the-moment thing. He fell out of a two-storey window. It wouldn't have looked so bad, but he was in the nude.' Milly says he's a different person now, and cringes at his past behaviour, but I'm laughing, and enthralled. I don't hear stories like this from my colleagues at university. It feeds something in me. Something feral.

The news turns to the flooding in Canterbury, and we hope Steph is getting home okay. The bridge is out at Porters Pass.

'Even I'd have trouble crossing that,' says Milly.

I ask Milly why he went to prison.

'Being stupid.' He was caught drunk driving three times. He was in one prison, but then, 'At the 2000 millennium when the whole world was going to explode, they moved us to mainstream.' While he was there, he built a new steelworks, got to work making a fence. 'I had 22 guys working for me by the time I left prison.' When he got out, he hermitised. 'Didn't like seeing anyone.' He travelled, though, spent two years on the road. 'I had itchy feet. Couldn't shut a door for a long time.' In his shorts, Swanndri and boots, 'knife in my pocket, everything I owned in my backpack.' He was happy.

We're quiet, watching the fire, Maz and I well into our second bottle of red wine.

'You remember Dave?' he says.

'I went out with him, didn't I?' I respond. I haven't thought about him for years.

'Yeah, you did.'

'What happened to him?'

'Dead, hopefully. He was a ratbag son of a bitch.'

He starts telling stories. When Maz and I became Christians some of our old friends ran amok in ways we can only imagine. We're aghast to hear that most of the guys from his class at school are now either 'alkies or dead'.

Some of them died in motorcycle accidents – a slippery manhole cover in the rain, a collision with a truck. But most of them are suicides. A shotgun, an overdose, gas. Surely this is more than statistically normal. We wonder what was going on with our school, our communities.

'I think nine out of ten of us had shit upbringings,' says Milly. 'Your primary relationship is with yourself. Good advice is to be kind, be gentle.'

Maz and I talk about our jobs for a bit, how much time we spend working, how focused we are on our careers. I explain how this isn't a holiday for me, I'm working, interviewing a few people, taking lots of notes, writing about it all. It's an aspect of my work that I love, that grounds me and energises me. Milly nods. I'm used to people telling me I work too much, or care too much about my work, but he gets it.

'Work keeps the sadness away. Keeps your mind off the silly shit.'

This seems the right time for me to bring up climate change. 'Do you think about it?' I ask Milly.

'It scares the crap out of me,' he says, 'but I don't look into it too closely because it fucken breaks my heart.' He's focusing on local initiatives, though, doing what he can. They won't solve climate change, but they will help. He's started doing some volunteer work with the Clean Streams programme.

'I'm a water keeper. My future job is to ensure the water is clean,' he says.

'What water? What do you mean?' I ask.

'*The* water. I'm a Pisces.'

I'm writing furiously in my notebook and exclaim as I come to the end of it and start writing on a random scrap of paper. Maz and Milly seem happy with my attention.

'It feels great to be finally acknowledged,' says Maz.

'We're animals in your field study.'

Once we're back on the couches, over a late-night cup of tea, I mention that my great-great-grandmother Marian – from my maternal line, she's the one whose mitochondrial DNA I carry – was born in Tasmania, but her mother came from England, her father from Scotland.

'Where?' asks Maz. 'Do you think that's where our ancestral connection is?'

We've decided we're related, or knew each other in a past life, though I'm not entirely serious about that possibility. Milly remembers our distinctive friendship from our school days. 'You were inseparable. You guys were just so goddamned comfortable with each other.'

'We're like an old married couple,' says Maz.

I say I'm off to bed to read, and hold up my book, Clare Moleta's *Unsheltered* about a mother searching for her child in an apocalyptic post-climate-change world. We talk about books for a while, about how it's hard to get them up here in Karamea, or even Westport.

'I want to get a book about how to make a suit of armour,' says Milly. 'I'm going to make one before I die.'

'Why?' asks Maz.

'Because they're fucking cool. I've got my eye on a sixteenth-century design.'

Milly can fix things, build things, grow things. My mother used to get dreamy eyes when she'd talk about a future after the collapse of civilisation, when we'd live primitive lives, off the land, like our forebears did. I didn't understand what would be so great about this future she was looking forward to – how would we fare without modern medicine? – but I grew up with the sense that there was something not quite right with the life we had now.

I'd forgotten about my doctor plan by the time I was ready to go to university, and studied earth sciences instead. Or maybe after the end of the Cold War we all felt more optimistic about the future, though I think I just wasn't ready to move away from my hometown, go away to university.

How did I end up being a writer and academic who writes about climate change? It's been a series of small steps rather than any overarching ambition that has led me to where I am now. I still find myself wondering if there's a future world where it will be essential to be useful, and whether I will have anything to offer. Would it be enough to be the archivist – to collect the stories, keep them safe? I can garden too. Forage. Cook. Teach.

But I want to believe in a different future. One that builds on the good things in the civilisation we now have, but is fairer, sustainable, less anxiety-inducing.

We start the next day with fat whitebait fritters served with a wodge of lime. Milly got the whitebait in exchange for fixing someone's motorbike. He doesn't go whitebaiting – 'It's too political' – and has a philosophy that 'you don't pay for wild food'. The children won't eat whitebait – 'because of the eyes' says the youngest – so there are more for us. We meet Steph, who came in late after an overnight trip to Christchurch. She's American, a relatively new arrival, but talks about how Karamea has changed over the last two years. Prices for a block that was $100,000 ten years ago have doubled. There's a 'mall', they say. A new build, a complex with four shops. Things were full-on this summer, they say. Karamea was packed with New Zealand tourists.

'Assholes,' says Milly. 'Don't even say hello when you pass them in the street.'

After breakfast, Maz and I walk around town, check out the 'mall'. We look at a real-estate window display, showing houses and blocks of land in Karamea and surrounds, but nearly all of them are *SOLD*. Along the main street there's a Four Square, a general store, an information centre, a tourist shop for booking tours, a café. Many of the shops have window signs with photos of Red Bands and *Please remove gumboots*. A noticeboard advertises a family-friendly winter solstice event at Ōpārara Reserve; acupuncture, Reiki and massage; opening hours for the local library and rubbish dump.

While Maz heads into Lena's hair salon, which is in the mall, I go next door to order coffees. I head back to the salon and sit on the couch while Lena cuts Maz's hair, then a woman with a German accent – similar to Lena's – brings the coffee to us in shiny white cups that we can return later. Lena is an excellent hairdresser – people come up from Westport to see her, she says. As she's finishing the cut, Lena's next client arrives. Before we leave, I ask Lena and her client, a blond woman who is a decade or so older than us, if they're going to have the vaccine.

'I'm not sure, I need to do some more research,' says Lena. 'Not just a tiny pamphlet written for anyone to understand.' She's smart. She wants to know more. The client has made up her mind. She's already had a 'dressing down' from her sister-in-law, who is a nurse. So she'll be getting it.

'Someone is coming up soon to vaccinate the whole community,' she says.

⁓

Last time I taught my science writing workshop, half my students wrote about their crippling eco-anxiety. But the anxiety of Maz and my teenage years was not ecological – climate change, biodiversity loss, pandemics – it was technological: nuclear annihilation. In 1981, the Doomsday Clock sat at four minutes to midnight. In 1984, citing 'the accelerating nuclear arms race and the almost complete breakdown of communication between the superpowers', the Bulletin of the Atomic Scientists moved the Doomsday Clock forward to three minutes to midnight, the closest it had been since 1953.

We spent our teenage years expecting a northern-hemisphere nuclear war between the United States and the Soviet Union. For a while it seemed we would be okay in New Zealand, a nation of back-to-the-landers going it alone after most of the northern hemisphere had been obliterated, but there was always the chance we'd be hit – every time a US nuclear-powered or armed warship entered our harbours we became a target. After 1983, when we saw the movie *The Day After* on a school trip, we knew that even if we weren't hit, there could be a nuclear winter to look forward to, when particles in the atmosphere would block the sunlight and we'd all starve to death because we wouldn't be able to grow food. Sometimes we wondered whether that would be worse than dying of radiation poisoning, which we knew about from the Raymond Briggs comic *When the Wind Blows*. In this book, which innocent children all through the 1980s read, attracted by the illustrations and the author – he also wrote *Fungus the Bogey Man* – follows a sweet old couple, Hilda and Jim, living in the country outside London. They survive a nuclear strike in a homemade shelter – a triangular structure made from a few repurposed doors and a pile of cushions – but over the next few days they grow progressively more unwell, with headaches, fatigue, vomiting and diarrhoea, hair loss, bleeding gums. No wonder we sought obliteration of our minds in marijuana, alcohol, Christianity.

The 1987 book *New Zealand After Nuclear War* was an academic study of the environmental, social and economic impacts on New Zealand of a major nuclear war in the northern hemisphere but it was the 1986 book *A Nuclear Survival Manual for New Zealanders*, by Brian Hildreth, that tapped into darker stuff and envisioned a full societal collapse. Hildreth's book has chapters on 'Surgical midwifery', 'A nuclear disaster first aid kit' and 'Plant medicine' but it's not a collaborative world he envisions; it sounds violent and patriarchal, with instructions on how and where to hide your food cache, how to set up a decoy camp, care of women and children. What with these books, the movie *Sleeping Dogs*, the TV series *Survivors*, the End Times tribulations of Christianity, the counter-culture back-to-the-land movement, there were multiple narratives telling me and my generation the story that *this* would be over soon, there would come a time when existing structures and civilisation would fall apart. This was either a wonderful opportunity to start again and create a better world, or it was really fucking scary.

Was it all this end-of-the-world talk that gave Maz and me a taste for living on the edge – for the thrill of getting close to danger but evading it, as well as a hankering for a kind of cooperative self-sufficiency, a way of surviving without the structures and systems that shaped and controlled our world? In February 2020, as news of the coronavirus outbreak started competing for headlines with the Australian bush fires, Maz booked an international trip, responding to an invitation to go sailing with a friend in the Virgin Islands. As she arrived at Auckland airport, the day the first case of COVID-19 was detected in New Zealand, she looked up at an orange sky, smelt the smoke in the air, and texted me 'Fuck it, I'm going.' She went sailing, drank rum with her friends, forgot about the pandemic. On her 22-hour journey home, she engaged with Twitter updates of international borders closing, increasingly alarmed texts from me, and departure boards filled with cancelled flights. She started texting me photos of herself, wide-eyed and masked, as she successfully made her way through each airport – St Thomas, Miami, Houston – heading closer to home. No one else was wearing masks, she told me. She landed two days before New Zealand went into lockdown. 'Thank fuck, I'm home,' she texted me, then hit the shops to stock up on cat food, loo paper and wine, to supplement the food supplies her husband had been laying down in the two weeks she was away.

I had a more cautious response to early news of the pandemic. As COVID-19 case numbers began to rise, I hit the garden, frustrated by the media pundits claiming the virus was no worse than seasonal flu, saying 'the cure should not be worse than the disease'. I angry-weeded one of my garden boxes, clearing a fall of cabbage palm fronds and pulling out a thick spread of chickweed and grasses, to create space around one perpetual rocket plant and three small parsley seedlings. I watered them with a solution of 'worm tea' and water, spoke some encouraging words, then went inside to call my daughter who was five weeks into her first year away at university, to tell her it might be a good idea to pack her bags and try to get on a flight home.

On my last outing before lockdown, I drove along the coast road to the landscape supplies centre where I filled the car boot with firewood. Up the hill, at the university that looks over the harbour, I retrieved my desktop computer and ergonomic chair. Downtown I bought a 20-kilogram sack of flour and a jar of yeast to add to the supplies Jonathan had been gathering over the past few weeks.

On the first day of lockdown I cleared the rest of the garden boxes and planted all the seeds I could find – half-empty packets of radish, radicchio, carrots and cavolo nero. The world had gone mad, but I was home with my family and felt deeply, profoundly, content. One of the anxieties of climate change – unlike the nuclear arms race – is that those of us in high carbon-emitting countries are all complicit. It felt good to be travelling less, doing less, reducing our carbon footprints.

'I feel like I've been waiting for this all my life,' said Jonathan, another Gen Xer like me and Maz, born in 1967. At last, the bad thing had happened. The bad thing had happened and we were prepared and were going to be okay. There were no zombies in the streets, we weren't holed up in a bunker armed with guns and surrounded by booby traps. Society hadn't collapsed. Rather than fighting and hoarding, people were queuing at two-metre intervals outside supermarkets, putting teddy bears in their windows to entertain passing children, heeding the Prime Minister's exhortations to be kind to each other. At home, we settled into a routine of bread baking, gardening, working from home and watching the Al Jazeera news in the evenings. When I had a hankering to rewatch the 1970s TV show *Survivors* – set in a post-apocalyptic England after a deadly pandemic had wiped out 98 per cent of humanity – Jonathan obliged with a box set of DVDs from high on a shelf in our home office.

'We're going to have to start again,' says Abby, when she makes contact with another group of survivors. 'We're going to have to relearn all the old crafts. Right from the beginning.' I found it exhilarating the first time I watched it, in the 1970s, and it was exhilarating to watch again in 2020. But this romantic idea of starting again, with fewer people, a new fairer system, doesn't last long. 'Our civilisation had the technology to land a man on the moon. But as individuals we don't even have the skill to make an iron spearhead,' says Abby. 'We are less practical than Iron Age man.'

Maybe Christianity – or the intense End Times Christianity Maz and I got involved in – was a distraction from the real problems of the world. Similar, perhaps, to how in 2021, some people, including Christian groups, are criticising the government for mask and vaccine mandates when the real global issue is government and businesses dragging the chain on climate change action.

I look back on myself and Maz as teenagers and see our need to be rescued from the anxieties of our age. We swapped fighting racism, sexism, and nuclear war for fighting Satan, infidels, and Worldliness. What a waste of energy. But when Maz and I talk about it, we agree that Christianity was more than just a distraction. There was something real there, something we can't explain away.

7

The Upper Room

Maz and I were excited and curious about going to the Upper Room, but nervous too, so my other best friend Adam came along. I'd known Adam since I was five and had been through primary and now high school with him. He was brought up Catholic and now went to a charismatic Catholic youth group; he was one of our first friends to start talking to us about Jesus, Christianity, living a meaningful life.

Behind the Assembly of God church, at one end of the carpark, we found a group of people standing around smoking next to cars and motorbikes. The motley crowd, mostly dressed in black, looked just like people hanging outside a band venue. We walked through them, then up external stairs to a brightly lit room filled with teenagers, where we were relieved to find the boys who'd invited us. They welcomed us with smiles and awkward hugs that connected arms but not bodies. They introduced us to everyone. People were talking and laughing and there was an air of anticipation.

After a while, an imposing woman with curly grey hair said it was time to get started, and everyone formed a circle. One boy picked up a guitar and everyone started singing.

Some of the songs were really silly and had actions. We felt like dorks but we sang along while we giggled.

Father Abraham, had many sons
Many sons had Father Abraham
I am one of them, and so are you
So let's all praise the Lord – *right arm* . . .

The songs gradually got more serious and more complicated, and someone handed us pieces of paper with the words. Then people closed their eyes and raised their arms, and at the end of one really serious song the guitar player kept going while people started speaking in tongues and praying.

'Hallelujah!'

'Praise the lord!'

'Thank you Jesus, my lord and saviour.'

Then the woman, who had a smaller man with comb-over and moustache by her side, asked people who wanted 'a touch from God' to line up. Some people walked straight up, and others went because she called them.

'Close your eyes, raise your hands, start thanking God,' she said to the first person in the line, a skinny boy about our age, wearing jeans and black T-shirt. When his eyes were closed, his arms in the air, and his mouth moving in prayer – or was he speaking in tongues? – she touched him lightly on the forehead and he fell backwards into the arms of the tall boys, our new friends, who lowered him gently to the ground. We were captivated and watched, intently, trying to determine what was going on – was she pushing them, were they faking it, or was something mystical happening?

Once there was a row of bodies on the floor, ecstatic looks on their faces, the woman said, 'Has anyone here not been baptised in the Holy Spirit?'

Maz and I looked at each other. Had we? Then we stepped forward and confirmed we had been born again and were ready to receive the Holy Spirit. We stood before her and closed our eyes while a group of Christians gathered around us, praising God, eyes closed, speaking in tongues, one hand raised to the ceiling and the other reaching out towards us. With hearts beating fast, and hands starting to shake, we closed our eyes and raised our hands, started moving our mouths, praising God, saying, 'thank you Jesus, thank you Jesus' while the voices around us got louder and louder, and the intensity built up and at the push of her hands on our foreheads we fell backwards and then we were lying on the floor and we were speaking in tongues too, and we felt fizzy and elated and peaceful and excited all at the same time. When I opened my eyes for a peek, there was the row of slain bodies, stretched out on the floor, eyes closed and mouths moving in prayer, speaking in tongues, thanking God.

While this was happening, Adam sat leaning up against the wall, with his legs crossed, watching.

After the singing and the prayer, we felt exhausted but happy. Everything seemed somehow different, like something fundamental had changed inside us; we had been transformed, and something now connected us, at some deep, deep level, with the rest of the Upper Room community. We stood around drinking instant coffee and eating gingernuts. The grey-haired woman came over to us and said, 'Where's the Holy Spirit?'

'Inside me,' said Maz. I nodded: same.

Then an intense look came over the woman's face, and she nodded towards Adam.

'Has he done transcendental meditation?'

'I dunno,' said Maz. 'Maybe?' Like us, he'd tried all sorts of things.

'I thought so,' she said, and started muttering to the man with the moustache while glaring at Adam.

We later learned that transcendental meditation was Satanic, a dangerous practice that could invite demons into your body. We went quiet, impressed that she could sense this much just from looking at Adam, but also feeling protective of our friend.

Adam didn't come to the Upper Room again, and after that we didn't hang out with the charismatic Catholics anymore. We prayed about it and decided that we would move from our little church with the kind, bearded pastor to the big church in Lower Hutt, where all the young people went.

～

We had Bible Class on Tuesdays, the Upper Room on Fridays, and Church on Sundays. We weren't lukewarm anymore.

Going to the Upper Room became the highlight of our weeks. We sang songs, praised God, and had fellowship over coffee. Some of the people who hung around the carpark, smoking and looking at the ground, were backslidden Christians. Sometimes they would come into the Upper Room, looking pensive or even crying, and the grey-haired woman would minister to them, and lay hands on them, and they would come back to the Lord. Mostly, though, they just waited until we were done and then we would all go to McDonald's – the Christians, the backslidden Christians, the ones who hadn't been saved yet. Some of the backsliders had motorbikes, tattoos, leather jackets. Sometimes we tried to talk to them, but they were usually wasted. We felt bad for Jesus that they had turned their backs on Him. And we praised God that He had been so merciful to us. But we were also scared, scared that the demons that had a hold on the backsliders would reach out and try and get a grip on us. We worried that there might be cracks in our faith, and became intent on building up our armour of God, which we learned comprised a belt of truth, a breastplate of righteousness, a shield of faith, a helmet of salvation and the sword of the Spirit.

At the Upper Room one night, the grey-haired woman told us all that – contrary to what some of our new friends had said – there was no such thing as a 'spirit of smoking', or a 'demon of drinking', rather there were just people making bad choices. After going to the Upper Room for several weeks, we made friends with some new boys who smoked cigarettes and listened to Christian heavy metal, like Petra and the Resurrection Band. We hassled our new friends, telling them their bodies were a temple for the Holy Spirit and they should stop smoking. We didn't smoke anymore because we had the Holy Spirit living inside us. We had so much joy overflowing in our hearts that our classmates at school, even the ones who had been brought up as Christians, couldn't handle it and would roll their eyes when we sang songs and talked about Jesus in the common room, and we wondered if they were demon-possessed because they weren't proper Christians. Maybe their demons couldn't cope with being so close to the Holy Spirit who was living inside us.

Sometimes the boys would give us a ride home. They liked coming to our places, because our parents let us sit around the kitchen table, drinking coffee and eating biscuits, until late. Once we'd shared our testimonies, our stories of how we got saved, we talked about our lives before, in the World. About how bad we used to be.

Then the boys told us about the End Times prophecies in the Book of Revelation. Some of the evangelists said the newly introduced barcodes and EFTPOS cards were signs that the End Times were already upon us.

We got grocery items out of the cupboard so we could examine the barcodes. There were three long bars, all the same length, and they were the sixes, which was evidence that the barcodes were the beginning of the 666 system. The Book of Revelation said that 666 was 'the number of the beast', and this all meant the Antichrist was coming soon.

We learned that some people said the Antichrist was Henry Kissinger, because of his support for the European Union. Others said it was the Pope, because of his support for the World Council of Churches. Both organisations were part of the move towards One World Government and the coming global financial collapse, which would be followed by the introduction of the Mark of the Beast – a 'mark' on your forehead or right hand without which you could not buy or sell. It was all prophesised in the Bible, and Nostradamus had worked out when it was going to happen, and that time was now.

Our friends were excited when I said I had some American money – and when I got an American dollar bill out of my bedroom, they showed us how the eye inside the pyramid was a sign of the Illuminati, a secret society whose members are in control of Everything.

After the boys went home, we lay in bed and wondered who God had picked out for us. We looked forward to getting married and being good wives. But we worried that maybe there wasn't time. The End Times were upon us, and anyway, none of the Christian boys seemed to be looking at us like that.

Some weeks, instead of going to the Upper Room, we caught the train into town and went witnessing in Manners Mall, telling people about the Lord. It was scary at first, but we learned to identify people who were open and seeking – young people looking sad, lonely, sitting by themselves. We would approach them gently, with a big smile on our faces, then tell them all about Jesus and invite them to come to Church with us. Sometimes they would let us lead them through the Sinner's Prayer and they would be born again right there and then. We also went to the Scripture Union bookshop to buy Amy Grant LPs, stickers saying *Jesus is alive*, and badges with *Praise the Lord* and *It's really great to be a Christian* written on them in bright bold letters. Sometimes, when we tried telling people about Jesus, we got persecuted. But we knew this was a test, and we prayed to Jesus to make our faith stronger. At home, our parents sometimes persecuted us too – for leaving a mess in the kitchen, or for being closed-minded. We tried not to get a persecution complex.

One time, when we went into town, we met some people from the Upper Room with a guitar and tambourines, singing songs of praise outside of the Manners Street McDonald's, the same McDonald's we were once kicked out of when drunk on cider with the punks. We stood watching for a while. Then, when they beckoned, we joined in – one shy, one confident – singing along, clapping our hands and banging a tambourine. Then we saw one of our mothers come past after a meal out with her workmates. We always held a glimmer of hope we could one day lead our mothers to the Lord, but this would not be that day. She'd had a few wines, and when she started dancing and clapping and singing along in a loud tuneless voice while her friends clapped and laughed, we felt humiliated

105

and angry. Then we felt bad that we felt angry.

Sometimes our mothers tried to have serious talks with us, in which it was clear they didn't respect our Christianity. They told us about their friends who were Quakers, or Anglicans, who did good works and volunteered in the peace movement. But we'd learned that churches that focused on good works were full of dead Christians, and it didn't matter how many good works you did; in John 3:3 Jesus said that 'no one can see the kingdom of God unless they are born again.'

But apart from John 3:3, which everyone agreed on, we were confused about which bits of the Bible you were meant to take literally and which bits not so much. Our Christian friends, many of whom were brought up Catholic, told us that many of the rules set out in the Old Testament – like 'An eye for an eye and a tooth for a tooth' – were superseded by things Jesus said in the New Testament, like 'Turn the other cheek'. But when we asked why some things in the New Testament were not taken literally, like 'If your foot causes you to sin, cut it off', things were less clear. When we got confused about things like that, we just gave it up to God and prayed for faith to accept the things we didn't understand. And anyway, we loved reading the Bible. We were fascinated by the history told in Genesis and Exodus, we giggled over the sexiness of Song of Solomon, and we connected deeply with Ecclesiastes. 'Everything is meaningless!' we reminded each other when we were feeling upset.

The most exciting part of going to church on Sundays was praising God. Each week, we would let the Holy Spirit move us to speak in tongues, quietly at first, while we got the hang of it, and then more confidently. Sometimes we'd talk afterwards about the word sounds people made when they were speaking in tongues and judge whether it was real or not. We sometimes even wondered if we were making it up ourselves, but how could we be when it was accompanied by feelings of such joy, transcendence, ecstasy.

Sometimes, in the middle of the praise, someone's voice would rise above the others. They would call out, in an earnest and meaningful voice, a string of sounds, and then, after a pause, someone would follow in English, translating the prophecy into messages that sounded like they were from the Bible: 'Behold! I say unto you the gates of heaven will open to the faithful and I will prepare for you a bed of roses and a drink of nectar.' It was exhilarating, but sometimes my brain got in the way of my heart. I'd find myself analysing the complexity of the English translation compared with the simplicity of the speaking-in-tongues sounds and wondering how they could be the same message.

But we liked listening to the sermons too, because our pastor was young and kind and funny, and told us we were all 'God's little masterpieces'. Before the 1984 election, our pastor said he wouldn't tell us who to vote for, then gave a sermon about Jesus caring for the poor, which we took as support for voting Labour. We were smug when we told our mothers about the sermon; they were convinced our church was bigoted and right-wing, and we hoped this might help them think differently.

One Sunday our pastor said he was a bit worried about the young people in the church, that some of us were taking things too far, and he explained that it was okay to read more widely than just the Bible and Christian texts, and we didn't have to listen to only Christian music. All things that are good, he explained, come from God. All the good in the world. Classical music, literary masterpieces, great art. We were relieved to hear this because we had finished reading the

Bible – all of it – and now we started reading books from the olden days, when people weren't so sinful. I especially liked *The Lord of the Rings*, *I Capture the Castle* and anything by Jane Austen.

When there were plans for a church production, we joined the church choir – one confident and accomplished soprano and one shy and middling alto – and spent hours sewing our choir dresses, transforming the acres of peach-coloured polycotton into flouncy sleeves, full skirts and bodice ruffles. When our parents saw us in our choir dresses they laughed until tears rolled down their faces. The more we were persecuted at home, the more we began to feel that our family, our *real* family, was our church family.

By now we were regularly invited to do things with the Christians. Some Saturday nights we took guitars and blankets through the pines and down the track to the dam where we sang songs to Jesus by moonlight. Lying on our backs, looking up at the stars, we wondered if any of the boys whose arms we'd made momentary contact with on the blankets, or who'd lent us their jackets because we were cold, would be the One.

Now that we were in the choir and our hair had started to grow back, we felt like we were beginning to fit in, but sometimes things still reminded us we were different. When Billy and Amanda got married at the church, everyone was invited. We didn't know what to wear to a wedding but we knew that this was an occasion for dressing up. Maybe if the boys saw us in dresses, they would feel romantic and realise their feelings for us, we thought. We looked through our wardrobes but the only dresses we had were cut short and held together with safety pins so we searched through the dress-up box: riffling through the rayon tea-dresses we used to buy at op shops, the Victorian lace shoulders-to-knees combination underwear that used to belong to Maz's great-aunt Maudie, and the accessories we'd raided from our grandmothers' wardrobes. We dressed, compromising our goal of looking feminine with our attempts to make each other laugh, and turned up in assemblages of florals and prints, heels and hats, embroidered gloves, and long coats in some kind of anachronistic pantomime of middle-aged ladies.

When we got to the church, and the tall boys we so wanted to like us greeted us at the door in their grey usher suits, we realised we had made a mistake.

'Hello, boys,' we said, in an attempt at a 1940s glamour voice.

They gaped at us, speechless, then ushered us to some seats near the back of the church.

Embarrassed, we whispered to each other and scrutinised the other girls' clothes. What did Normal girls wear to a wedding? As the ceremony progressed, we whispered about the horror of the wedding night, having sex for the first time with someone you'd barely even kissed. We knew it was what God wanted but thought it was a risky proposition.

'You know what we should do when we get married? Dye our pubes.'

'Yeah – Bahama Blue and Ultraviolet.'

'Imagine what our husbands would say when they saw it.'

'Unless they wanted to do it in the dark?'

'Imagine going to all that effort and they didn't even notice.'

While Billy and Amanda read out their vows we shook with silent laughter.

Until Amanda vowed to 'love, honour *and obey*' her new husband. Obey? Wasn't that something from the olden days?

We stopped smiling then, and looked at each other wide-eyed, with raised eyebrows, then looked around the room to see if anyone else was feeling uncomfortable.

~

We drove to summer camp with a boy called Paul, in his Sunbeam Rapier, as part of a convoy of cars from the church. In Bulls we had to stop because Kev's Holden was overheating. In Tūrangi we had to stop because Sam's Cortina had a blocked fuel lead.

'Satan is out to get us,' said Paul, while we waited for all the cars to be roadworthy. When we got to Taupō, after 11 hours, we laid our sleeping bags under some trees by the lake and crashed out for the night.

At the Tauranga Christian camp the next afternoon, we pitched our massive old canvas army tent and invited five other girls from church to share it with us. The boys from church were in the tent next to us. We were surprised that some of the girls in our tent had sexy underwear and were giggling and flirting with the boys in the next tent.

That evening, at the first service, the Australian pastor taught us that shyness was a form of pride, and pride was a sin. So I learned to smile a big friendly smile, showing my teeth, and say 'Hi!' and it seemed to encourage people to talk to me more. Our friend used to say, 'If only Rebecca had more confidence, she'd have everything going for her', and after that first sermon I had everything going for me because now I had confidence as well as Jesus.

Maz and I had invited an old school friend, who had moved to Taupō, to join us, but after one day at camp she said the food was repulsive and the meetings were boring, and left. Well, we didn't like the food at camp either, but we were grateful that we lived in a country where everyone had enough to eat. Maybe she just couldn't handle it. She was really into Nina Hagen when we were at school together, and we thought it was possible she was demon-possessed from listening to NunSexMonkRock. So we prayed for our friend, and when she came back, we told her this was her opportunity to know God. Why didn't she try and seek Him? Finally, she decided to open her heart to God and see what happened.

That evening, there was an altar call for anyone who wanted baptism in the Holy Spirit, or just a new touch from God. Just about everyone went up, including our friend from Taupō. That night, she was saved. We were so happy, we spent ages just praising God. Everyone hugged and some cried. But after a while she got annoyed with us for being bossy, telling her what to do, and because Maz was saying, 'Raise your hands, start thanking God . . .'

One afternoon we went to Mount Maunganui where there were hundreds of tanned bodies, all over the beach, like in California. We witnessed to some people on the beach then we saw someone we had gone to school with, so went to sit on the sand with him.

'What are you doing for New Year's Eve?' he asked us.

'Praising God,' we said, and told him we were Christians now. When his mates heard us talking, using the words God and Jesus and Church and Bible, they gathered around to listen to us. We shared with one guy for about half an hour. He sat quietly with wide-open eyes and a smile, and he seemed open to what we were saying. He said we were beautiful. No one had said that to us for a long time, but we figured that now that our hair had grown again we looked a bit more normal. But we also thought that maybe he was really stoned.

That night there was a concert and everyone from our church got up and sang a couple of songs. At about 1am, someone put a film on. We went to bed about halfway through, as we were bored and tired and the boys we liked weren't even there. When we left, to walk back to our tent, we saw the boys with two new girls from church, sitting in the dark and talking quietly and giggling. These new girls had permed hair, make-up, tight jeans and flirty eyes. It made us wonder when God was going to find us nice Christian boyfriends. But it also made us confused. Weren't we meant to be pure? The older, married Christians told us that if we had a boyfriend, it was best not to spend time alone – 'Too tempting!' they would say with a knowing laugh – and before we were engaged to be married we shouldn't be doing anything more than holding hands, or exchanging a goodnight peck on the cheek. Passionate kissing, they said, could lead to other things, experiences that would violate a person's purity, and what if you didn't end up marrying the person you kissed? Do you really want to be pashing someone else's future spouse? Even though we had no boyfriends to tempt us, we lay awake wondering what was right, worrying why it was all so confusing, and what we were doing wrong.

The next day, at a service about God's plan for our lives, the pastor told us there was a ministry for each of us. At the end, when there was an altar call for anyone who God had spoken to about their ministry, we were moved to go up. It was the first time we had really cried before God – at church, that is; we cried all the time at home – and we felt that we really wanted to do something for God. We wanted to be useful in his Kingdom. We didn't want to spend the rest of our lives just trying to keep in line with God's work; we wanted to have a real ministry in which to serve him. We had been crying out to God about it, right from our hearts, but for ages we hadn't been listening to Him telling us to have patience. We had only been thinking about all the other people who understood God's calling on their life. It was hard trying to cope with bad feelings as a Christian. Normally, out in the world, we would just get drunk and wreck stuff. But as Christians we could only talk and pray and give it up to God.

The praise-and-prayer session went on and on. After the service, Maz and I discovered that we'd each had the same message from God. It made our spines tingle and our hearts fill with joy. God had told each of us that our futures were in His hands, and all we could do was get on with living our lives and praising God. For the first time in ages, we felt real peace about our futures.

At supper that night, we sat at the end of the room and just talked. We didn't worry about who was or wasn't looking at us, or sitting with us, or whether we'd ever get a boyfriend, or what God's plan for us was. At least our church didn't have nuns, we figured. It would happen eventually.

While everyone else travelled back in convoy towards Wellington, we caught the bus to Whakātane and stayed at an aunty's house. While the olds and their friends had a session on the balcony, we lay on the waterbed and read Mills & Boon books and tried not to inhale any of the pungent smoke. We couldn't afford the intercity bus, so we decided to hitch home. We helped ourselves to a bag of peaches and a jar of almonds to sustain us, threw on our packs and hit the road. We got a good ride straight through to Taupō, where we met some of the Christians who were camping at Acacia Bay. We had lunch with them then they gave us a ride to Tūrangi – me in a car, and Maz on the back of Mark's 1100cc motorbike. The Christians decided the Lord's protection might not be enough to see us home, so Mark offered to accompany us to Wellington. We didn't know if having a tall leather-clad man on a motorbike a few metres away from us scared people off but it was hard to get a ride. We finally managed to get a lift to Waiouru with a Mormon guy in a Fairmont Ghia. He was really eager to hear our testimonies and we had a pretty good share. Mark followed on the bike. At Waiouru, where we had a munchies stop, it started raining. We waited in the rain until we got a ride with some old guy to just out of Taihape. Then some skinheads with beards and tattoos on their knees – one of them quietly adjusted the rip on his jeans to show his – offered us a ride to Palmerston North.

We declined the offer, thought we might check out what Taihape was like on a Friday night. By the time we'd eaten our burgers and chips, we were beginning to wonder if we'd make it to Wellington. We walked out of Taihape and along State Highway 1, but it was dark, we were a couple of hundred kilometres from home, and there were hardly any cars. So we accepted an offer of a place to stay from some farm boys who lived in a small town just south of Taihape. We were getting cold. Maz was wearing jeans, a jersey and Mark's black leather motorbike jacket but I had bare legs, my black leather skirt – if the Christian boys were going to keep wearing their leather jackets, then I wasn't going to give up my favourite piece of clothing – reaching only halfway to my knees.

'Do all Christian girls wear leathers?' the farm boys asked.

'Yep!' we replied. Might as well. We liked messing with stereotypes.

They drove us up a dirt road, past the farmhouse, to a large shed bathed in moonlight. With Mark as our guard, and angels watching over us, we spent the night in a woolshed in Utiku, our sleeping bags stretched over soft and smelly bales of wool.

8

Limestone

A fat-bodied bird, dark with a white front, is bouncing around our feet, hopping up to stand on a wrench in an open toolbox, then pecking invisible morsels from the pale dirt around us. It's a kakaruwai, a South Island bush robin.

'They'll come sit on your gumboots if you put crumbs on them,' says one of the Department of Conservation workers. A dog wearing an orange Conservation Dogs New Zealand vest ignores the robin, who has completely captivated me. The dog is certified to sniff out great spotted kiwi and whio, says his handler. I take photos and videos of the dog and the robin to send to my children once we're back in range.

We've stopped in the Ōpārara Basin, parked next to a group of DOC workers, after driving north through farmland in Milly's truck. Karamea used to be the biggest swamp on the West Coast, Milly tells us. There were flax mills here a hundred years ago, exporting flax fibre from the estuary of the Karamea River for use in ropes, twines and lashings. But once the harakeke was gone, the wetlands were drained, the land flipped and farmed, and the grazing animals moved in. We all agree it's a tragedy. Wetlands are an important habitat for plants and birdlife, but they're also like giant sponges, protecting surrounding land from flooding when there are high rainfalls or storm surges from the ocean.

At one end of the car park, an information board features a map of the Ōpārara Valley and various tracks – to the Mirror Tarn, the Ōpārara Arch, the Moira Gate Arch – and stunning photos of the geology and biota, with captions boasting of a 'rich and vigorous fauna, ranging from worm-eating snails to tree-dwelling bats'. But the board also tells us that whio and kākā are endangered and numbers of weka and roa are declining in the Ōpārara. Effective predator control is important. One panel tells the story of the South Island kōkako, and the last confirmed recording of its haunting call in 1967. Some people think this bird, which nests and feeds close to the ground, has survived predation by rats and stoats and still lives in the forests.

Milly tells us a mate of his heard a kōkako in the bush on the Nydia Saddle in the Marlborough Sounds. 'He's not an . . . ornicologist, what is it?' he asks.

'Ornithologist?' I offer. Milly is silent for a moment.

'This would be the place for one,' he says.

Milly spent years working in the Marlborough Sounds to help build a new backpackers lodge as part of a project to keep 1080 out of Nydia Bay. Income from the lodge paid for traps, boots and food. I push him on the 1080. He's 'not some conspiracy theorist', he says, but using traps rather than 1080, a biodegradable but controversial poison used in pellets to kill possums and rats, gave people jobs, something to do. But while he doesn't think 1080 is part of a plot to exterminate the human race, he does have some concerns. 'What's it doing to the invertebrates? What's it doing to the microorganisms in the soil?'

Other panels feature the now extinct moa, other forest

birds, and giant moss, liverwort and snail species.

The arches that bring the tourists here are made from a limestone sandwiched between granite bedrock and a cap of mudstone. This fossiliferous rock is formed from tiny skeletons and shells of marine animals that lived and died 30 million years ago, falling to the ocean floor before being crushed, compacted, uplifted, and exposed at tourist attractions and quarries.

We set off on the mossy, stony trail, lined with dripping ferns and high trees. The river, on our right as we walk, is an orange brown, the colour of the billy tea we drank at the flames, darkened with tannin from fallen beech trees. It's cold and damp and fallen trees sprout an abundance of fungi, mosses, grasses and worts. It's not raining right now but there's a drip, drip, drip of past rain making its way through the canopy and the sound of the river. I ask Milly questions as we go, about the birds I can hear, the plants I can see, but he has an oblique way of answering my questions that seems to challenge my way of looking at the bush.

'What kind of tree is that?' I ask. The tree has a moss-covered silver trunk and a proliferation of small, rounded green leaves on its reaching branches.

'It's a fucken beautiful tree,' Milly replies.

'Is it silver beech?'

'Yes,' he says quietly, then points out a young lancewood, a straight and thin trunk with a cluster of long, spiky leaves at the top.

'They make incredible walking sticks,' he says. 'Very light and very strong. They grow up to where the moa can't reach them, then they branch out with the succulent leaves.' I know that moa, and hōkioi – a giant eagle that was the moa's only non-human predator – used to live in this area. There are local businesses offering tours to visit limestone caves that hold entire intact moa skeletons.

'This was the main hunting ground of the hōkioi,' he says.

I want to talk more about the plants and critters that live here – the anemone mosses on the side of the path, the black pīwakawaka that is flitting among the trees, the predator traps I can glimpse in the undergrowth – but Milly and Maz have started talking about their recent relationship break-ups.

'I know if I start crying I won't stop for a very long time,' Milly is saying. I hang back, walking on my own, jotting in my notebook, letting them talk. I'm feeling dreamy and content, enjoying the exercise and cold air, absorbed in the life of the forest. When I catch up with them Maz is telling Milly about my writing project.

'In the book I come across as a badass with a heart of gold,' she's saying with a smile and a wink. I agree, then repeat 'badass with a heart of gold' while ostentatiously writing it down.

Milly takes us on a short side track then stops and I think we've reached the edge of a cliff. In front of us is a big nothing, a white cloud over a dark void, but as I approach the edge I realise the void before us is a brown lake, a tarn, and the cloud is a layer of mist hovering above the water – the Mirror Tarn. The mist clears as we stand by the lake, revealing blue skies, beech trees reflected in the water, and grasses, reeds, and ferns growing around the shore. It's quiet.

'Unbothered by humans,' says Milly. 'The beauty of a place like this is you have to make an effort to come and see it. If we had some more time, I'd like to take you up the Blue Duck.'

'What's up the Blue Duck?' I ask.

He pauses, then says, 'Blue ducks. They're like a cartoon version of a duck.'

I know them as whio, one of Aotearoa's rarest birds, a blue-grey duck with a bronze-coloured front and yellow eyes and beak.

As we walk on, they continue talking about relationships, and how the end of an intimate relationship doesn't mean a failure of that relationship. 'Different relationships bring different things into your life at different times,' says Milly.

Why me and Maz now, I wonder? The main things she's bringing into my life are adventure and laughter. Googly laughter that makes me feel weak and tingly. Belly laughter that makes me double over, tears streaming down my face, and after which my abdominal muscles hurt. I wonder what I do for her? I think she knows she can talk to me about anything. I won't judge her and I won't share her secrets. I may offer a reality check now and then. 'Thanks for the call,' she often says after we've talked on the phone, usually about her worries about her job, her separation, her life choices. Without realising she's a lifeline for me too. I've been so focused on work and family for so many years I'd almost forgotten how to have friends, how to be a friend, and I feel like I'm getting something back that I'd lost. No one gets me the way that she does.

'The right things happen at the right time. But it very seldom feels like it at the time,' says Milly. I don't really agree. I don't think there's a guiding force and I think a lot of shit stuff happens for no reason at all. But it's comforting to imagine there is a purpose, a reason. Sometimes, for my sanity, I need to believe there is some purpose to my life, to *all* life, and I latch on to it, in different ways. I pray, meditate, give thanks, I'm just not sure to who or what. As for what's guiding my life, after reading a book called *Creative Visualization* in my tumultuous post-Christian years, I started writing lists of what

I wanted in my life. By my early forties I had everything on the list. Travel, babies, trips to Antarctica, books published, a PhD. I had a bit of a crisis when I got there, had to figure out what came next. An academic career? Check. But what now? I feel like I've been trying to learn how to just live, be here now, enjoy what I have rather than striving for more. The pandemic has helped with that, I think with a wry smile. I've actually enjoyed the lockdowns, found them a lifesaver, they let me slow down and tap into something I needed.

The lockdowns also brought me and Maz closer. After a friendship where she always seemed a bit more suited to the world than I was, more able to just get on with things, we found ourselves in a world to which I was better adapted. While she was struggling with being stuck in her house, unable to go to work, socialise with friends, be out in the busy world, I was happily gardening, writing, conducting all my meetings online. We started talking most days. I wasn't her only lifeline – as an extrovert she needed a network – but it brought us closer and changed the dynamic between us. It made me feel I'd made up for the years of being a shit friend. I apologised to her once, a year or so before the pandemic. We were sitting on bean bags on the grass outside a waterfront bar and I hadn't seen her for ages and then she told me she'd invited along a couple of other friends. I cried. And I told her I knew I'd been a shit friend, so caught up in my work and parenting that I had no time for anyone else, but I explained that she was still my number one, my BFF. When she invited other people to our infrequent catchups, I felt like I was just one of the gang, one of her Wellington friends. I cried, and we talked until we laughed, and she put off the other friends. Since then, we've been closer. We have been friends long enough to have a few scars – some times in our wild post-Christian years that we can't talk about without getting emotional, blamey, jealous – but we're old enough, and have each spent enough time in therapy, to know this is a special kind of friendship.

We follow the track as it takes us higher, above the river. Another fat black robin on the path distracts me for a while. There are dead trees, topped by Cyclone Ita. There were evacuations in town back then. 'That's when you realise how tight knit a community is. You see enemies working side by side,' says Milly. 'We spent the best part of a year cleaning up after that one, cutting up trees, clearing fences.'

When we reach the Moira Gate Arch, we stop. The magnificent limestone structure sweeps over the dark orange water, dripping with brilliant green ferns. It's cold, and we huddle in our coats and take photos of each other with the arch as backdrop.

Back in Karamea, Milly hands me a plastic bread bag containing a packed lunch – two white bread whitebait sandwiches, two apples and a big bag of scroggin – and we load our gear into the car to hit the road again. We take the main road out of Karamea and wave to two police officers parked outside the school.

We stop on the way out of town to look at a sculpture of a person with moko, taiaha, cloak, riding on a hōkioi. With a wingspan of up to three metres, it was the biggest eagle to have ever lived. A plaque, acknowledging Te Rūnanga o Ngāti Waewae – the West Coast hapū with its marae at Arahura just north of Hokitika – marks *one of the many stops along the West Coast pounamu pathway from Tuhuroa (Farewell Spit) to Piopiotahi (Fiordland in the south).*

We head south – there's no other way to go – back through Waimangaroa, Ngakawau, Westport and Charleston, which was once a thriving gold-mining town, now with placards advertising glow-worm caves, a rainforest train and underwater rafting.

We eat Milly's lunch on the beach, then continue south, past stands of tī kōuka, ponga, nīkau. We drive by fields containing grazing pūkeko and derelict machinery, remnants of anchors, boilers and tractors. We pass a house on stilts. The whole lower storey is space; I can imagine the sea washing underneath and wonder if they built it that way with sea-level rise in mind, to be ready.

At Punakaiki, we splash out on a hotel by the beach. It's probably time for a shower and a change of clothes, but we can't shake the feral – we're beginning to tap into something elemental. There's a phrase in the back of my mind that the Australian writer Charlotte Wood said, about feral women artists, and I make a note to look it up.

Our upstairs room has a balcony that looks out over the wild beach and ocean, a prime spot for an end-of-the-day drink. Maz opens a packet of chips and I get out the gin bottle. The tonic I bought in Westport has an old-style metal bottle cap and I can't find a bottle opener so I hand it to Maz, the engineer, to sort out.

'You want me to open it with a knife or something?'

'Actually I thought you could do it with your eye socket,' I say, certain she would have learned some tricks in engineering school, outback Australia, down a mine somewhere. Maz looks around, sees the brass plate that holds the door's security chain, and with one hand on the bottle and the other on the plate, gives an expert flick of the wrist then hands me the open bottle, tonic water foaming over the side. We have a moment of surprise – it worked! – before Maz snorts then slides down the door and onto the floor in a fit of giggles, just as I realise her vigorous flick has removed a whole inch of glass and left a jagged edge.

I take our glasses and the broken bottle onto the balcony to watch the sunset. Maz follows with the chips and pulls the sliding door closed to keep the warmth inside. There's no one about – we're one of three cars parked in a large complex – so we light up on the balcony. Across the carpark is a sandy beach with waves chiming in from the Tasman Sea. At the horizon, a row of cumulus clouds tower before a sun setting yolky and bright below apricot sky. We watch the tī kōuka and harakeke become silhouettes. When it's time for dinner, I stand up and wait for Maz to open the sliding door. She leans up against it, applies pressure to the handle but nothing happens. She slides to the floor.

'What are you doing?'

'I've locked us out,' she says.

Now I join her on the floor. 'Fuck.'

When we're tired from laughing, I start to formulate a plan.

'How's your upper body strength?' I ask, which just starts Maz laughing again.

'No, seriously,' I say, raising my voice in excitement. 'If you can lower me over the balcony, and onto the tarmac, then I can go to reception and tell them I've lost my keys . . . but I would need to keep a straight face – what if they could smell what we've been smoking?'

'It's the West Coast, no one will give a shit.' Maz laughs until her eyes form black slits and mascara runs down her face.

While I'm talking through variations on the plan, Maz tries the door again and this time it opens. She had been pushing it rather than pulling it. We decide we're too far gone for the hotel restaurant, so take to our beds, with a bottle of red wine, a second packet of chips and a block of chocolate on the bedside tables between us.

At the Punakaiki Reserve there are thin layers of limestone, stacked on top of each other like pancakes, that have been eroded into pillars and caves, tunnels and blowholes. At the edges of the walking track are vertical limestone bluffs. Signs saying *Danger – do not go beyond barrier signs* sit next to orange lifebuoys ready to throw into the churning water below. We run along the asphalt pathway in the weak morning sun, gasping at the sights we don't have time to stop for. I have an appointment to get to – an interview in Hokitika with a well-known climate change denier.

We drive south past Barrytown, where Milly recommended a Make Your Own Knife place. *You get to forge your own blade from red hot steel, and complete your knife with a native timber handle, brass bolsters and pins*, says the website. We note its promise of *you'll amaze yourself!* And *Lunch is provided, usually toasted sandwiches* and make a plan to return. We continue on past the Strongman Mine Memorial, which commemorates a devastating 1967 coal mine explosion. The government-run mine was being lax about safety regulations, Maz tells me, and an incorrectly fired charge – intended to blast some coal into small enough pieces to remove from the tunnel – instead created a fireball that killed 19 men. As we follow the highway inland, we drive past more evidence of mining history – rusting metal pylons, coal buckets, steam engines left there so long they've transitioned from junk to treasure. The skies are clear and there is snow on the mountains to the south. We pass families of pūkeko in the fields, paradise ducks in pairs.

The road takes us inland and through Rūnanga, a town whose name means 'meeting place' and where Europeans built a settlement of wooden houses in the late nineteenth century, homes for the coal miners. It's a cold morning and the old wooden houses are already sending coal smoke out of their brick chimneys, contaminating the low mist hanging in the valley. We're getting close to the Grey River. We cross at Greymouth – noting the muddy limestone cliffs exposed on the south side of the river – and drive past the Tai Poutini Polytechnic where Maz used to teach health and safety and risk management to miners and quarriers. We continue south through Paroa, where my grandmother was born. Through Arahura where my grandmother's best school friend, who was Ngāti Waewae, had her marae. And over the mighty Arahura, the pounamu river, its origin high in the Alps, past signs which say it's part of a river restoration project, an effort to restore native vegetation to the river banks.

I'm going to interview the Westland mayor about coal and climate change, and I need to step into role. At Kūmara Junction, just across the Taramakau River, I order a three-shot flat white with oat milk. We listen to Nouvelle Vague's cover of 'Too Drunk to Fuck'. I put a talisman in my bag, a copy of Rebecca Solnit's *Men Explain Things to Me*. I've learned a lot from the scientists I've interviewed – about Antarctica, climate change, biodiversity and more. I've also got used to asking scientists questions to which I already know the answer, to get quotes I can use in my writing. And I've learned to be tactful, generous, submissive even, with the occasional prickly or controversial scientist, such as Richard Dawkins and Lawrence Krauss. If I want people to talk to me, to tell me things they've not told every other writer who's interviewed them, I need them to trust me, even like me a little bit. I'm okay with that, because the interviews are about them, not me. I've also used my work to tap into the concerns of scientists and other thinkers about the future, like British cosmologist Lord Martin Rees, who in 2002 wagered that, by 2020, 'an instance of bioerror or bioterror will have killed a million people'. Or environmental journalist Mark Lynas, who talked about planetary boundaries, and said that while 'nature can still give us a good kicking', humans were still in charge. 'It's up to us what temperature the planet is at,' he told me. 'It's up to us what the acidity of the ocean is; it's up to us what species survive on this planet. We have to close the cycle of the economy so that we're not constantly bringing resources in from the outside and then disposing of them.' When I talked to primatologist Jane Goodall, she had a more positive take on the future and a lot of faith in the younger generations. 'Change is possible, but we've got to do it quickly,' she said. 'I think we've only got a limited window of time to change attitudes so we can learn to live in harmony with nature and stop destroying it.'

I've interviewed hundreds of scientists. Now I'm keen to try out these techniques on the mayor of Westland. I want to talk to him because he claims the government's environmental policies are a greater threat to the West Coast than climate change. He says things in the media like 'I love coal'. I know some of my colleagues are frustrated with the things he says, and I've heard that his staff are challenged by his stance on climate change. The Westland District Council's draft 2021–2031 plan acknowledges 'the reality of climate change' and 'the need for urgent and decisive action' to reduce carbon emissions, but it also states that anything we do here in New Zealand won't make much of a difference. I got an interview with the mayor by telling him I want to hear *his* perspective – I'm not going to try and convince him of the reality of climate change; I want to hear what he believes and why, and try to understand where he's coming from.

I get a warm feeling as we pass the Hokitika town limits, a clear view of the snow-covered Southern Alps, Ka Tiritiri-o-te-Moana, out our front window. I text my mother a photo – she'll recognise where I am, even without a message. As the place where my mother's mother grew up, the word Hokitika has always kind of meant 'home' even if I never lived here. We drive into town, the war memorial clock tower marking the town centre, and Maz drops me outside the council buildings. The buildings are on the site of the old Hokitika Police Camp and Gaol, established in the 1860s, not long after my great-great-grandparents arrived in New Zealand, Edward Garland from Sydney and Marian Milne from Hobart, each one generation away from their homeland of England. A display says the camp had *eight-foot-high corrugated iron fences, two lock-up cells, accommodation for the men, a watchhouse keepers house and a stable for the horses.* Today, the site is occupied by a three-storey modern building with lots of big windows.

'Climate change is just part of life,' Bruce Smith is saying. 'When I went to school in Westport in the 1950s, in the winter, the puddles were always frozen, and now they're not. And our summer used to start in December, and now it starts around about the second week of January. So climate's changing, and you know, I think there's an awful lot of emotion and a lot of nonsense out there, that I read, but that's the same in most aspects of life, isn't it?'

We're sitting in black leather chairs. The mayor's desk is piled with papers and around the room are a hi-vis vest on a hat rack, models of a Hitachi digger and an Airforce Iroquois helicopter, a few cores of granite, a miner's lantern, and lots of photographs – mostly landscapes and grandchildren. Bruce is a fifth-generation coaster – born in Greymouth, schooled in Westport, and a Hokitika resident for the last five decades. He's coming up 69, and loving the job, he tells me. I try to win a bit of cred by telling him I'm on a road trip with a friend who was general manager at Bathurst for a while. I also tell him that I work with climate scientists and write about climate change. Then I cut right to the chase.

'Some of the things I've read in the media describe you as a climate change denier, and I wondered is that true? Is that how you see yourself?'

'No, not at all,' he says. This is not the answer I had expected, and I ask, if he acknowledges the climate is changing, *why* he thinks it is changing.

'Because the location of the earth to the sun is changing,' he says. I let him continue. He's saying things that I've heard before – climate change is 'natural'; it's not really getting any hotter. It's true that climate changes in response to long-term cyclical changes in Earth's orbit around the sun, and the angle of its tilt, that impact the amount of solar radiation reaching the Earth's surface. These Milankovitch cycles have been responsible for past ice ages and interglacial warm periods. But it's been well established that the rapid warming happening today is not because of these cycles. If it was just the Milankovitch cycles affecting things, we'd be heading into a cooling period.

He carries on talking about the local weather. 'We haven't had a decent flood since 1982, which was the big one in Franz Josef, which washed out a few walls.' The tornadoes are a new thing, he says. 'We had a couple two or three years ago.' But the Coast has always been a place of dramatic weather. 'We still get 12 metres of rain a year at the Cropp, nothing changes there.'

The Cropp River, which starts high on the western slopes

of the Southern Alps, has one of the highest annual rainfalls in the world, the mountains high and steep enough to cause moist landmasses from across the Tasman Sea to rise and cool and precipitate, but not high enough for snow. They get three and a half metres annual rainfall at Hokitika, and five metres at the top end of Kahurangi, he says. 'Nothing's changed there. As for sea-level rise, the measurements at Jackson Bay and Greymouth have remained pretty much unchanged for 40 years.'

In South Westland, I know from my work with the NZ SeaRise programme, local tectonic uplift is working to cancel out much of the recent sea-level rise, but sea-level rise is accelerating as the Antarctic and Greenland ice sheets melt. The West Coast won't be able to rely on local uplift for much longer.

'So, do you worry about climate change and sea-level rise?'

'No.'

'The scientists I work with would say that the climate is changing because of the carbon dioxide that we've been putting into the atmosphere since the start of the industrial revolution –'

'No,' he says. 'It contributes. But I don't accept that at all. Every volcano that goes off would have more impact than all the human intervention there's ever been. So, it's part of it, but it's not the significant part, and it's not the only part.'

I try to follow what he's saying. I don't have time to educate him on the nature of science, and I promised I wouldn't push my views on him. But I'm thinking that massive volcanic eruptions, which spew particles high up into the stratosphere, can shield the earth from solar radiation and lead to short periods of cooling – but these sorts of eruptions are rare. It's wrong to say they have more of an impact than greenhouse gas emissions. He's got a valid point, but he's following it to a false conclusion. I'm keen to find out where he gets his information from. What makes him believe what he believes?

'Oh, I live here,' he says. 'I get up every day. I walk to the beach every day. In the 70s and 80s, I hunted in the bush for 20 years. Observation. And, you know, the thing I see these days is the scientific community is very, very keen on investigating all sorts of things as long as they can get funding for it. I mean, we see it with the Alpine Fault – when the funding rounds open, all of a sudden the media's full of Alpine Fault information.' When it comes to climate change, he says, you typically get five news articles a day that mention it. 'It all looks a bit orchestrated to me.'

'Orchestrated, how do you mean?'

'Well, it's not realistic to have five articles every day, on anything.'

'Yeah, but what is *orchestrated*?'

I can't pin him down about who's making it happen, who's conducting the orchestra, and he says, 'I prefer to deal with reality and the things that I can see.'

I ask him what he's reading to inform this perspective.

'I sit back and look at it and have to make a judgement based on my own experience,' he says, 'which is what I do.'

Does he have many detractors?

'No. I have a Facebook page that has 30-odd-thousand followers,' he says. 'If we walked down the street, you won't find anyone talking about climate change.'

'So, just reading between the lines a bit there, it sounds like you have a – is it a bit of mistrust of scientists?'

'No, not at all. No, science is great, but it's only part of the package. And science that is there for the purpose of accessing funding is a lot less valuable than any other sort of science.'

A common climate-change-denier talking point is that

climate change researchers are just following the money, taking advantage of the 'multi-billion-dollar Climate Change Industrial Complex'. It's true that funding – government or private sector – can help dictate the direction of scientific research, but again the mayor is taking a valid point and twisting it.

'It's not that I distrust them,' he says, about scientists. 'I always look and say, okay, who's paying the bill? And why are they paying the bill?'

It's a good question to ask, but I think government funding for climate-change research and geohazards is money spent to protect the people and the land.

Our conversation turns to science funding and an ongoing project to better understand the Alpine Fault – when it moved in the past, when and how it's likely to move again. Smith doesn't seem too keen on funding for that either.

'Every person who lives on the West Coast, *everyone*, knows that we live on a faultline. They know that most of us either live by the river, or we live by the sea, and we live here because we want to live here, because our parents and our grandparents lived here.' He says he has a pile of reports on his desk about the Alpine Fault, and he's read them, but if there's an earthquake, he jokes, 'It will be them that kills me,' rather than anything else.

'And it won't alter the fact that we're going to get an eight or a nine, maybe in the next 50 years.' He's right about the statistics, and I'm surprised he's not more concerned. 'So to me, it's about resilience, from a practical point of view. It's about preparation.'

This is the way Maz talked when she lived on the Coast. You live here by understanding the risk and being prepared, she would say. She had an evacuation kit handy, then just got on with her life.

I bring the conversation back to sea-level rise, point out that that planet is going to experience 30 centimetres of sea-level rise by 2050, and it will be even higher in some areas of the coast where land is slowly subsiding. But Bruce brings up the earthquake argument. 'You've got to take into account in every Alpine Fault earthquake, our ground here's moved up between one metre and four metres.' He points to the raised river terraces we can see out the window. 'All exactly the same distance apart. All have risen exactly the same height.' Every 300 years, he reckons, an earthquake raises the land around three or four metres.

'So that would be your expectation for what would happen here? That the Alpine Fault moves and lifts the town up?'

'Well, we're being told by the scientists in their bucketloads that we are at year 300 of a 300-year cycle. We're told that the other side of the faultline moves around about eight metres to the north, and we're told that on this side of the island here, we rise between one metre and four. It should be in every report. Either that, or it should be ignored and thrown away, because it's either true – it's backed by science – or it's not.'

I want to talk about sea-level rise throughout the wider district.

'Ever since I was born, and going back years, the sea comes in and goes out. There's always been areas that have been badly eroded. The beachfront hotel, the water used to be right in underneath there. And that was about, you know, early 1900s. At the moment it's going out in lots of places. It's like living next to a river. Every so often there's a flood that's a ripper, does a hang of a lot of damage. This town, before it had its rock protection, was damaged hugely by floods. So, once again, it's preparation.'

'So are you just saying that it's a changeable coastline?'

'It most certainly is.' He's got a bach in Punakaiki, he says, where 'Māori ovens' that were once on the beachfront are now 200 metres inland. 'So, you know, the sea comes in, the sea goes out, and it's part of life.'

I ask him if he has any explanation for the dramatic retreat of the Fox and Franz Josef glaciers. 'Well, the glaciers are receding at the present time,' he concedes. 'In my life, I've seen them come down just as fast, but they are receding, and I think they'll keep going backwards for a long time. That's my view.' A lot of it relates to rainfall, he says. 'If you go back ten years, and there's huge rainfalls, then the glacier will come down. A lot of it is related to that. I don't know the full ins and outs, but looking at the glaciers, they're heading backwards.' He owns a helicopter and is familiar with the landscape from the air. 'We have 26 glaciers on the Coast here. I've flown around most of them over the years. And they're beautiful. And definitely seeing change. And it's a bit of a nuisance, because it means that "Glacier Country" has got to look at re-branding itself.' The alternative, he suggests, is putting a tourist gondola up the Franz Josef Glacier.

He's a lively conversationalist and I'm enjoying talking to him. But it's bringing on a deep weariness inside me.

I wonder why it is that he trusts his own observations – *looking at the glaciers, they're heading backwards* – so much more than a scientific report that can tell him the Franz Josef Glacier has retreated three kilometres in the last 100 years. My sense is that he has some reasonable talking points, but he takes them to illogical conclusions. If I find him halfway convincing, what about people who don't have a science background, don't read the IPCC reports, don't talk to scientists every week?

'We are going down to the glaciers tomorrow,' I say. 'It's been about 20 years since I was last there.'

'Oh, you'll notice the difference,' he says.

The Franz Josef Glacier has been retreating steadily since 2008 and the walking track, which follows the river valley up to the glacier terminus and was washed out by a landslide in 2019, doesn't extend as far as it used to, making the distance between lookout and glacier even longer. After the river shifted from the north side of the valley to the south, the track was not rebuilt, which he describes as a political decision. 'They need to put a bulldozer in, like they've done historically, and flick it across to the other side.' Visiting the glaciers used to be a part of tourism and local life here, he says. 'It was almost an annual visit for our family, to walk up and have a look.' Today, it sounds like the lookout is kilometres from the glacier and the only way for an intimate glacier experience is by helicopter.

I decide it's time to talk about coal – to me, the topics have a causal relationship: the melting glacier is a direct connection; the planet is warming because we're burning fossils fuels, and coal is responsible for nearly half of the extra carbon in the atmosphere. Maz totally agrees with that, even if she has a bit of a fondness for coal. I ask Bruce if he really said 'I love coal' in an interview, and he confirms that he did. '*Why* do you love coal?' I ask.

He explains how the extractive industries – first gold, then coal and timber – have been the backbone of the West Coast's economy. The Stockton coal, he says, 'is the best steel-making coal in the world, and it's critical to employment, critical to New Zealand's balance of payments, fabulous stuff.' I'm familiar with this argument by now, and it seems fair enough, until alternatives are found – but what about the shit coal, the coal mined from down in the Grey Valley, coal used locally for heat or energy production? He brings up the case of the local dairy factory, now owned by the Chinese company Yili, which burns about $50 million of coal a year.

'And the reality is, there isn't sufficient electricity for them to convert,' he says. 'Biowaste will never work. Hasn't got a hope, or so the scientists say. And the practical person would say it as well.'

At this point I can't help but have a little giggle. The idea that scientists are somehow not 'practical' people.

'There's nothing that can convert energy to heat like coal can.'

He's right, and that's why coal has played such an important role in getting us to where we are now. Which, from where I'm sitting, is not that great a place to be. While I'm brooding about the state of the world, Bruce starts talking about carbon capture, a technological solution whereby carbon is removed before coal's combustion gases are released into the atmosphere from their stacks. But hold on, I interrupt. 'If carbon dioxide isn't the problem, then why would we need to capture the carbon?'

Because it's good for horticulture, he says, and outlines a system where the captured carbon is directed to the crops. I'm not so sure I buy this.

I want to finish by talking directly and honestly about climate change. 'I'm someone who worries about climate change a lot. I lie awake at night and worry about it,' I tell him.

'You poor thing,' he says.

At this point, I start to feel pissed off but I keep my cool and stand my ground.

'I worry about the future for my kids. So what would you say to people who worry about climate change?'

'You know, there is so much rubbish in the media and across

social media. There is so much rubbish that it's no surprise that you're worried at night. It's no surprise that youth suicide continues to increase and mental health issues continue to increase, and it's because a lot of people are terrified. I'm not one of them. And most of the people I know are not one of them. I don't know anyone who doesn't sleep at night because of climate change.'

So we agree that there's rubbish and disinformation in the media, but I think my solid news is his fake news, and vice versa.

I remind him that we're heading for 30 to 40 centimetres of sea-level rise by 2050 and ask how the Coast is preparing for that.

For a coastal region, he says, 'There's a reality if the sea's in your back door, you've got to shift. And if you're going to shift, you may as well shift to somewhere higher. Nothing clever about it, it's just common sense.' He points out that the towns around here – Westport, Greymouth, and Hokitika – are all beside rivers and on the coast. If Greymouth hadn't built its flood protection wall in the late 1980s, he says, it wouldn't be here. 'They'd be catching flounder there again.' So he acknowledges the need for flood protection. Hokitika has a rock wall along the beach, and there's more beach protection to come. 'Money will follow logical paths,' he says. People looking to build new houses, or invest in infrastructure, will look at the risks. 'Will you build on the Alpine Fault? Well, probably you won't. Are you going to build right by the sea? No.'

I need to circle back to climate change, get his stance clear in my head. 'So I just want to get it clear, you do think climate change is happening, is real?'

'Oh yes, it is, but it's nothing new.'

I press him again on where he learned his argument about

natural cycles causing climate change. 'I've read things,' he says.

I want to know what things. Where did he find them? What are his sources?

'I read all sorts of things,' he says. 'I mean, if you look round the world, the reason there's seashells on tops of mountains is because of climate change. The reason that there's the skeletons of boats in the middle of deserts, it's because of climate change. Nothing's changed. We've just got climate change. We've always had it, except probably in the past it's been much more extreme than what we've got now.'

The reality is that our climate has been stable for 10,000 years. And seashells on tops of mountains is more about tectonic uplift than climate change. I tell him about the scientists I work with, like Tim Naish from the Antarctic Research Centre. Tim has studied Milankovitch cycles, and his clear conclusion is that the climate change that's happening today is not what would be expected from natural cycles but is caused by fossil fuel emissions from burning coal and oil.

'That's great,' says Bruce. 'It's good that he's done his research and he has an opinion. It's really important that he has an opinion, but it's different to mine. So it's of not much value to me.'

'But does the fact that he's a scientist carry weight for you?'

'Not at all,' he says. 'I don't put scientists on pedestals. I mean, I'm an electrician. Does that carry any weight?'

I'm frustrated now. He's not arguing in good faith. Does he really think his *opinion* is valid? Does he not understand data, evidence?

'But he spent decades doing this research,' I say. 'Does that not mean that he's a good source of knowledge?'

'Look, it's all about taking in what's out there, and placing credibility on what you read, and making your mind up based on your own observations. I'm not going to sit back and say Ted at Victoria University or whoever, he knows what he's talking about, therefore that's what I'll be thinking, it's just not going to happen.'

I ask again where he's getting his information from.

'Well, virtually every article I read comes from a scientist somewhere. They've all got different views. And so, at the end of the day, you're stuck with a whole lot of different views, but then for me it comes down to my observations of real things and practical things, and that's where I come from. I'm not the slightest bit interested in all of the hypothetical nonsense that goes on. I'm interested in what's happening around me, what's happening in my community.'

I'm focusing hard on what he's saying. What I would call conclusions based on data, he's calling 'hypothetical nonsense'. His focus is local – what's happening in his neighbourhood, community, region. 'Whether it's Alpine Faults, or cyclones or storms, or river floods, for me it's all about resilience and preparation,' he says. 'And the rest of it's – you know, half of it won't happen. You know, I feel sorry for the kids these days.'

I phrase it carefully, so as not to pigeonhole him as right-wing or conservative, and mention something I've read that identifies a difference between the people who rely on their own observations and experiences – what they can see around them – versus people who are more open to other ways of thinking.

'Or, there are people who are leaders and rely on their own abilities, and there are people who can't, and they need to take direction and guidance from someone else,' he responds.

'So, do you think – I don't want to put words into your mouth – but do you think there's a risk that people are just sort of being sheep-like and doing what they're told?'

'Absolutely. You're not putting words in my mouth; that's bang on. It's no good following the pack. You've got a responsibility to read what you can, understand what you can, have a look at your own surroundings, and then have a look at what's best for you, your children, your grandchildren, and that's pretty important. That's how I see it, you know? And it comes down to who you listen to. You know, do you want to go out and listen to a 16-year-old kid out there screaming and squawking? It's not something I'd do.'

'Is that a reference to Greta Thunberg?'

'Absolutely. What a joke. A friggen nutter.'

His talking points are all things I've heard before – right wing, anti-science, anti-elite, individualistic and misogynistic. It's pissing me off. I know I promised not to push my perspective on him, but I do feel like pushing back.

I ask about the vaccine. I'm wondering if there is some alignment between climate change denial and COVID-19 vaccine hesitancy; the disinformation machine is certainly working hard on both these fronts. But no, he's on board with the need for vaccination, says his photo was in the paper two days ago getting his second jab. I tell him that many of the people I've been asking on the way down the Coast have been saying they're not sure about the vaccine, or talking about the need to 'do more research'. I say that from what I know, a lot of the people who have unwarranted conspiracy theories about the vaccine are also climate change deniers. But he sees it's the other way around, that 'a lot of the conspiracy theory people are mad on climate change'. He calls them 'the nutter brigade'. To Bruce, the nutter brigade includes people doing anti-vaccination leaflet drops, people 'pushing climate change', and people talking about rising sea levels.

'What about 1080?'

This one is different, he says, with a 'very small and focused group that believe we're being poisoned'. He loves what's happening south of Hokitika in the Perth River Valley, where a predator-free programme is using 1080 to eliminate possums, rats and stoats. 'And of course we've got that natural barrier of the rivers and they can work their way up or down the coast going to the next river.'

He talks about snowfall. Every year, he says, 'I've taken a photo on the first of May off my veranda at home, where I get deer wandering through my place and all sorts of things. It's quite beautiful,' he says. 'Every year, the snow arrives two or three days before, or on the day, and it hasn't changed in 20 years. I don't think I've had a year that it hasn't been the same. Some years, you know, a bit more snow than others, but they're observations that I make personally.'

I figure that his 20 years of observations, one observation a year for 20 years, is 20 data points. The scientific papers he's rejecting would have hundreds, thousands – even millions, if you count the IPCC reports – of data points that go into concluding that humans are responsible for climate change. Why does he equate his observations, and his online readings, with genuine scientific research?

I'm fumbling with my voice recorder – the interview has gone on for so long that I either need to end the interview or start a new audio file – when he starts talking about 'the latte drinkers in Wellington'. 'You know, we're a coast where people are resilient. When something goes wrong from a natural point of view, the community gets together – you fix it, you adapt. But you won't do that in Wellington if there's an earthquake. Every person will be looking for someone else to fix their problem. We don't. We fix our own problems.' It comes from living in a dynamic environment, he says. He's got a point.

Coasters are famous for being resilient, nonconformist, and they do have a great sense of community. I'd like to be more like that. Maz always says that if something went wrong over here, you'd have ten people with diggers turning up to give you a hand. This interests me, and I want to talk more about community, but he's back to talking about Wellington.

'So I don't mean to pick on the latte drinkers, but oh God some of them are whackos.'

'I just had a latte on my way down,' I say.

'My wife drinks lattes and there's nothing wrong with it,' he says.

The interview's finished. I stand and admire the view from the big picture window. Across the road is a tourist shop, Mountain Jade, offering Ngāi Tahu's precious pounamu to tourists who aren't here anymore, kept away by the pandemic and Aotearoa's closed borders. Beyond a view of trees, fields and forested hills, are snow covered peaks. He points out Aoraki and Mount Tasman. 'There are days when it's beautiful blue skies, no clouds in the sky, and I sit here, and I have trouble working, it's so bloody beautiful, you know?'

I do know. I've enjoyed talking to him – it's been stimulating, a chance for a mental workout – but I wonder why he was cagey every time I asked where he got his information from. Is he hiding right-wing climate change denial web sites? Some QAnon conspiracy theories? A disgruntled scientist who's feeding him information? Or did he just make up his mind years ago and doesn't want to change?

He sends me down the stairs.

'Don't fall over or we'll have to fill out a form,' he says as I turn to wave goodbye.

~

In about 1920, my great-grandfather Anders, an immigrant from Finland who had settled on the Coast, was kicked in the chest by a horse. His injuries meant he had to lie down for two years – says the family story – and my great-grandmother Lydia – Marian and Edward's second child – had to earn to support the family. She bought a little shop in Hokitika and each day would walk there and back, three miles each way, to sell books and china.

I meet Maz just a few doors down from Lydia's shop, at the Hokitika Sandwich Company. The café is decorated with rustic corrugated iron, rough sawn wood, and coffee sacks, with customers served on a selection of mismatched vintage plates and cups. On the wall, a photo shows Revell Street in 1885, two-storey verandaed wooden buildings with a road for horses and carts. Lydia would have been ten, I think, still living in the inland mining town of Māori Creek with her parents and four siblings.

Maz and I share a winter veggie sandwich – grilled pumpkin and beetroot, walnuts, goat cheese, red onion, pesto, greens – then I walk up the road to find Lydia's shop.

I have a photograph of my mother here, from the 1980s, but I'm having trouble finding the shop window, the inset door. I cross the road and see the Preston's sign at the top of the building, the old shop frontages replaced by huge plate glass windows, the series of small shops now one big chain store.

When Lydia started working in the shop, her daughter Gertrude, always called Trudy or Gertie, had to leave school. Her three older siblings were either working or at university in Christchurch, and she had to run the household and care for a younger brother and two motherless cousins while her mother worked and her father recuperated. Years later, she trained as a dental nurse in Wellington, but she was in Hokitika one summer when she met my grandfather. Her brother Ted, on holiday from Canterbury University, brought home a man named Walter for a hot meal and a bath. Walter was having a summer holiday adventure, swagging around the South Island with a gun, a fishing rod and a bag of rice. Trudy moved to Christchurch after she married Walter, who was by now a French and English master at Christchurch Boys' High School. She had two children – my mother Ruth was born 14 years after my uncle Bill – and kept house, did the gardening, painted and exhibited landscapes, and read widely – French and English classics as well as the *New Zealand Woman's Weekly*. Her life was less adventurous than her ancestors' or her descendants'. Following the philosophy of my socialist grandfather, who didn't believe in spending money outside of the country, she never left New Zealand.

I need to buy a new notebook, and we find a shop with china and knick-knacks like Lydia used to sell. I pick up a tiny bowl with floral border and gold trim. *Friends are angels following you through life.* I show Maz, and we smile. It's cheesy but cute. I buy a notebook, and some gifts for my children, and she buys a harmonica.

We drive out of town past Westland Milk Products, the factory that goes through 50 million dollars of coal a year. But I can't find my phone. As a travelling companion, I'm constantly flustered. 'Where are my glasses?' 'Where's my phone?' 'Where's my pen?' I have to ask Maz to stop the car. 'We're going to have to go into the dementia unit together,' she says. I don't think it's dementia, but I do think there's something weird about my brain. I can hyper-focus enough to write a book but struggle with the everyday things I've learned to call executive functioning. Then there are the things I call shyness, introversion, anxiety. The online quizzes I take every few months, along with the accounts of neurodiverse people I follow on social media, suggest it's something I might want to get investigated.

I find my phone, and we continue on down Livingstone Street, where we stop so I can take a photo of the house where Anders and Lydia lived when they moved their family from the town of Kaniere into Hokitika.

As we continue south and inland, in search of the area my grandmother's family used to holiday and live in, we talk about how our lives today are unrecognisable from those of our grandmothers, great-grandmothers, great-great-grandmothers. Maz was married for 20 years but was able to choose not to have children. I made do with three, and survived a twin pregnancy that could have killed me. We've each worked in fields that used to be restricted to men. We're on a roadie together. We have a bottle of Reefton-made gin rolling around in the back of the car.

So why does my life feel so hard? I have a demanding job, three children, a manic dog, two elderly cats, a dilapidated house, a mortgage to pay. But I mostly love my work, and I love my family. Is it worrying about the future that makes it all feel so difficult? Makes it hard to enjoy the life I have?

—

Lake Kaniere, a name I first heard as 'Lake Canary' when my mother used to talk about it, is beautiful. The source of Hokitika's drinking water, it's now surrounded by scenic reserve. The deep lake fills a depression scoured by a glacier during the last ice age, 12,000 years ago, when ice caps covered

the alps and sea level was 120 metres lower. My grandmother
grew up in the small township of Kaniere, before the family
moved to Hokitika, and used to holiday by the lake with her
family, first in tents, and then in a lake house built by her
father, Anders. His sawmill was not far away, in Kokatahi.

Last time I came here, following Cyclone Ita, the ten-
minute 'easy, wheelchair accessible' bush walks my family and
I attempted had turned into exciting hour-long scrambles over
fallen trees. Today is sunny and calm and all the weather drama
is on the east coast. The lake is now lined with impressive
new houses, giant tī kōuka trees and boats in the front yards,
bush-clad hills behind. I'm looking for Hans Island, where
my teenaged grandmother used to row a clinker-built boat
her Finnish father made for her. She and her cousin Daisy
would pack a lunch and spend long summer days on the
island, swimming, exploring, playing. In 1988, a year after my
grandmother died, my mother rowed out in a dinghy to bury
her ashes on the island. A decade later, when she was about
the age I am now, she swam to the island. It was summer, the
weather was fine, but the lake is cold and the island is further
than it looks, and it pushed her to her limits. When I told her
of my plan to swim to the island – the lake's not much colder
in winter than in summer – 14°C compared to 16°C – she got
angry. 'You must not!' she said. Now she's texting Maz to tell
her not to let me get in the water.

Maz parks on a concrete boat ramp, the wheels nearly
in the water. The island is in front of us. A shag sits on a
dead branch over the water. There's a couple on a park bench
looking over the lake and some people on a jetty further along.
Clouds obscure the peaks of the Southern Alps to the south.
The water is still, with the faintest ripple.

While Maz plays harmonica scales I walk along the lake

front, looking for the best spot to enter the water, then come back, following the tune. This is as good a place as any. I open the doors on my side of the car and get changed between the front and back door, taking off my layers of jeans, jersey and T-shirt and pulling on a black bathing suit. My menopausal body doesn't feel the cold like it used to, and I wade in quickly and confidently, turning to give Maz a mad grin before I plunge in. I'm swimming. I can do this, I say to myself, but I feel my face contorted into an expression of anguish. Mentally, I'm up for it, but my body is reacting to the icy water, every muscle clenched, saying *no*. I last two minutes, swimming in a semicircle away from then back to the boat ramp. I float on my back a few more seconds. Say *Nanna* into the lake. I have no need to go further; I'm in the lake she swam in. I'm sure that molecules from her ashes have made their way into the water. It's gorgeous. I'll come back and swim to the island in summer. Perhaps when there are people with kayaks and dinghies nearby in case I need rescuing.

I text Mum a photo of me in my bathing suit, in the water, to keep her guessing for a while.

Once I'm dry, we follow a dirt road around the east side of the lake, crossing steep gravelled creeks, some completely dry, with mounds of boulders and gravel. Maz and I argue over whether the piles of gravel are mining tailings, glacial moraine, or rockfalls. Maz insists they're mostly mining related. 'It's an engineered landform, that's for sure,' she says about a dry channel I can't understand.

Next stop is Dorothy Falls, where my mother scattered my grandfather's ashes, but only after three failed attempts to put them in the Waimakariri River. Walter was the only survivor of a 1926 rafting accident that killed his two university friends and she'd thought there was something appropriate about taking his ashes to the river where he'd fished, rafted, tramped decades earlier. But she was turned back three times, by floods, road closures, a flat tyre – and it gave her the feeling that he really didn't want to be there.

At Dorothy Falls, where Walter proposed to Trudy, we walk down a short track to a rocky clearing. The water cascades down a series of drops, over moss and fern-covered boulders, to a pool below. I pick up a piece of granite to take home with me. Where is the granite from? Are these erratics? I wonder. Rocks transported by glaciers to places where they make no sense? Two tiny pīwakawaka flit around us, fluttering from branch to branch. My feet are cold and the swim in the icy water has given me a headache.

We continue around the lake towards Kokatahi. Past steep fern-lined banks and more disturbed ground. Over Granite Creek bridge. Past a gate that Maz reckons is a mine – 'reworking the tailings', she says. We leave the lake and cross the Styx River, the Alpine Fault, the Kokatahi River, and drive onto a broad floodplain.

During the Ōtira glaciation, which peaked 18,000 years ago, the Southern Alps were covered by an ice cap and a glacier flowed from the mountains to the sea. The land scoured out by the glacier is now flat, filled with river gravels, but interrupted by a few monadnocks, remnants of bedrock too high and too hard to be eroded by the glacier. Floodplain creation is not something that happens now. Now that we have houses and farms along the sides of our rivers we build stop banks, to keep the river in a tight channel, stop the periodic flooding that would add sediments to the plain. They don't always work. We drive on to Kokatahi, a rural settlement with large sections, a few houses and a pub. We pass Mill Road and I wonder if it has anything to do with my great-grandfather Anders.

We had planned to shoot some pool and drink some Speights at the Kokatahi pub, but next to the *Please remove all dirty footwear* sign is a sign saying *Sorry closed. Gone poaching.* We continue on along the floodplain, which is mostly farmland scattered with the occasional large podocarp.

Then Mum texts me. I reassure her that I'm fine, I didn't really swim to the island, and send a photo of my smiling face as proof. I tell her we've just driven through Kokatahi.

'Did you find the street named after your great-grandfather?' she texts.

Newly informed, we turn back and go looking for Hackells Mill Road. We reach it at a crossroads, so first we turn left and drive towards the river. It looks like no one much comes down here. There are tyre ruts in the dirt, but the middle and sides of the road are heavy with tall green grass. Farm paddocks are on one side, a stand of trees on the other. We drive slowly, so I can look out the window for any bits of West Coast rusting machinery that might be a piece of an old mill. We continue on until we come to a waterway that's either a stream or a farm ditch. The 'road' continues on the other side of the water, but there's not a proper ford and we both decide it's not a good idea to take the rental car through the water.

There's an old, rusted metal bridge-like thing across the water, so I get out of the car while Maz turns around. I explore as much as I can, but a gorse bush and a tangle of blackberries mean I can't get onto the bridge, which is rust red and patterned with pale green lichen. Maz is doing a five-point turn, backing onto some grass, forward across the dirt road, back into some mud. Back. Back. Back. The wheels are spinning but she's not moving anywhere. I come over to offer . . . support? Advice? Traction? She winds down the window and revs the car while giving me a gleefully manic grin. Black mud splatters all over the car's front doors, bonnet and windscreen, and my legs. The car doesn't move. I go googly. Collapse forward over a laugh that makes me weak in the tummy. But no, I really don't want this.

'Are you serious?' I ask. 'Come on, I'm not having this.'

She's looking at me and smirking, slowly giggling. 'Can you find something to put under the wheels?'

I look around for a piece of four-by-two. There are no pieces of four-by-two. From inside the car, she passes me a sturdy cardboard sandwich wrapper and a used napkin. I contemplate them then chuck them back on the passenger seat. 'Come on,' I say, imagining us stuck here for hours, waiting for rescue.

'Hold on, I'll try this.' She presses the "S" for sport button next to the transmission. She revs the engine, I put all the muscle I have into pushing the car, and – after a few anxious seconds – the car moves.

We drive back up the road, watched by Friesian cows, across the crossroads and back onto a paved road. There are mountains ahead of us. Paddocks on both sides. A few lone rimu, remnants of the forest felled by my great-grandfather.

Hackells Mill Road comes to an end with some farm buildings, a house, a barking Labrador. There are no old buildings or rusting machinery to suggest there was once a mill here or anywhere along the road. I'm guessing it would be wise to have the mill near the river – to use the water to transport the logs – though Anders was notable in these parts for being the first person to buy Leyland trucks to move the massive logs.

We're on the second round of the playlist now, and we're singing along to the Topp Twins: *We don't let anybody touch our brains / We won't ever, ever plug into the mains.* The song

149

energises me and I need it. I'm tired. After talking to the mayor, and chasing my family, I need some nature to reinvigorate me, so I suggest a detour to the Hokitika Gorge which is just 15 minutes away. When we arrive, a display board tells the story of the local geology, and features the Alpine Fault, which runs down the island from Marlborough to Fiordland and marks the boundary between the Australian and Pacific Plates. The board describes the 'monumental impact' of the fault on the West Coast landscape, with stress from the crunching and grinding of the two plates creating the 3700 metre high Southern Alps over the past 10 million years. Ngāi Tahu have a different story. In Nic Low's book *Uprising* he tells of Aoraki, first son of Raki (Rangi) the sky father and the ancestor from whom all Ngāi Tahu are descended, and of his brothers and Te Waka o Aoraki – the main island the upturned waka they travelled on. It describes them all perched on the upturned waka, turned to stone in the cold.

Weather and climate shapes this landscape too. The West Coast is the wettest part of Aotearoa, and the rain often falls in a high-intensity downpour. It used to be colder here. When the rain fell as snow and glaciers, over the last half a million years, carved jagged mountain peaks and scoured out U-shaped valleys. Now, rivers flow in the valleys carved by the glaciers, and vast moraines deposited by the glaciers form barriers that turn the rivers into lakes.

As we walk the trails, and along a boardwalk jutting from the cliff face, I look into the light blue water, milky with glacial dust, and think about how much I love understanding how the world works, and knowing about the geology, with its narratives that tell us what has happened here over hundreds, thousands, and millions of years.

I think about what the mayor said about observation and

evidence. I don't believe something because someone tells me about it or because everyone else believes it; I need evidence. I love stories, especially the ones used to encode the truth, but the things I trust the most are unambiguous, provable, can be reduced to a mathematical equation. My Christian experience was an inoculation against not just religion, but any sort of received dogma or group think. I'm also comfortable not understanding some things. I don't need answers to things that are unknowable.

At a high point of the trail there's a gate opening to steps that lead down to a sandy rocky cove on a bend in the river. As we stomp down the wooden steps a pīwakawaka flits about us, back and forth across the path. And then there's another one, and another. When we get to the river bank, I climb a pile of boulders and sit laughing as ten or more pīwakawaka fly around us, landing on rocks, zipping in front of our faces, jumping onto the sand to pick at sandflies. Pīwakawaka is one of the children of Tāne, god of the forest, a little warrior challenging us as we enter the forest. Pīwakawaka is also known as a messenger, bringing news of death from the gods to the people, giving us an opportunity to prepare. But pīwakawaka always make me happy. It's a bird that makes me think of my Nanna. After her cat Sooty died, Nanna would spend hours in her garden talking to the birds, and when she came to stay with us would be visited by fantails in our garden. I look at the pīwakawaka and I think of my Nanna and I laugh. I'm just happy to be here.

9

End Times

At the end of 1984 we finished school, but we weren't ready for university. What would we study? Maz's engineer father steered her towards a government cadetship that would give her salaried employment while she studied for a New Zealand Certificate of Engineering. I trawled the newspaper for jobs suited to my science-focused secondary school qualifications. Our first step out into the world, after being inseparable best friends through high school, saw us with jobs as public service technicians next door to each other, high on the hill in Kelburn. I worked for the Meteorological Service, and Maz had a job at the Seismological Observatory. She interpreted seismograph traces, while I calculated soil moisture saturation, and sometimes – when we'd run out of real work to do – we did the photocopying, punched cards for the computer, stuffed computer printouts into envelopes, or transferred numbers from one piece of paper onto another.

At the Meteorological Service, each day at 10:30am the other technicians – who were mostly men – walked down the stairs to the tearoom. Fifteen minutes later, the scientists – who were *all* men – would arrive. Once I confirmed it was not an enforced rule, I took my morning tea any time I wanted and sometimes Maz would join me from next door. Gradually we got to know the other technicians, as well as some of the scientists, the librarians – both women – and the male and female university graduates who were training to be meteorologists. At lunchtime we ate our packed lunches in the tearoom or sitting under a tree in the Botanical Gardens. Sometimes we would buy burgers from a place we called the Sleazejoint, in a poky little building beneath the cable car tracks. We'd talk about work, and church, and boys, and sometimes she'd whisper to me that there had been 'another bomb'. The seismograms she interpreted each morning included traces from a Rarotongan seismograph which picked up a distinctive pattern of ground shaking caused by France's underground nuclear explosions at Moruroa.

Now that we had jobs, we could start tithing. We officially joined the church, which involved signing some forms, making a pledge in front of the congregation and, as adults now, giving one-tenth of our income to the church, just like our pastor had told us we should.

~

We were all excited about the upcoming Barry Smith crusade. He had a new book out called *Second Warning*, because people hadn't paid enough attention to his first book, *Warning*. Our church friends had told us about Barry Smith, who claimed that New Zealand was going to be used as a test market for the 666 system, which was coming soon. It went something like this: following the collapse of the global monetary system, each person would be issued with a unique number – six digits to identify your birthdate, six digits to identify your location, and a six digit tax number. At first people would carry the number around on a plastic card, but then things would change, and it would be tattooed or lasered on people's right hand or forehead. Like it said in the Book of Revelation, no one would be able to buy or sell unless they had this number, which was the Mark of the Beast. Smith said that after a while things would get so bad that people who did not take the Mark and did not pledge allegiance to the new world leader would be put to death. We decided that we would rather die than take the Mark. And we would rather run and hide than die.

'Is Barry Smith a prophet?' we asked our pastor.

'Wait and see if the things he says come true,' he said.

But then our pastor moved away to a bigger church. The elders prayed about who should take his place, and God told them to appoint a man who was as old as our grandfathers. We already knew him as a church elder, who made us laugh when he used archaic words like 'helpmeet'. 'Just as Christ is the head of the Church, the man is the head of the family,' he would say, 'and the Bible is clear that God made Eve as a helpmeet for Adam.' When a man was having problems, and needed the church elders to pray for him, he would invite his helpmeet up to join the prayer, to support her husband.

The first service under our new pastor featured a praise session with soaring sopranos, electric guitars and hypnotic drums. The new pastor let us get really into it, and someone at the front of the church started handing out banners – fringed textiles on poles, embroidered with the word of God. A line of people started gathering to take a banner, but our pastor intervened. 'Only the men,' he called out, handing the colourful banners to the boys from our youth group, who held them high and ran around the hall, around the blocks of stackable chairs where we stood praising God, along the back wall where pamphlets and petitions were stacked, to the front stage where the musicians played. Why did you have to be a man or a boy to carry a banner around the church? It seemed to be an interpretation of something that was written in the Bible, but we didn't know whether to be outraged that we weren't allowed to do it too, or relieved because the boys looked kind of dorky.

Then our new pastor delivered a sermon that was different to anything our previous pastor had said. He told us that AIDS, a deadly disease that had now made its way from the United States to New Zealand, was the wrath of God on homosexuals. He asked us all to sign a petition against the homosexual law reform that would legalise sex between men. He quoted from the Old Testament book of Leviticus, which said that a man lying with another man was 'an abomination', and told us if we didn't sign the petition we were condoning a mortal sin.

At the end of the service we joined the queue at the back of the church and signed the petition, but with heavy hearts. If the Bible said homosexuality was a sin, and our church thought this was so important, we reasoned, we could 'hate the sin, but love the sinner'. The next week, after gay rights activists found out about the petition, they held placards and shouted outside the church while we were having our evening service. 'Satan is outside,' our pastor told us. We sang loudly and prayed that God would protect us from their persecution.

When we got home, we told our mothers what had happened, and that we had signed the petition, and they shouted at us too. They started listing the names of their friends who were gay and accused us of getting caught up in a bigoted church, of being brainwashed, of abandoning the values we'd been brought up with.

155

The next week we wrote letters to Fran Wilde, the Labour politician who was championing the Homosexual Law Reform Bill, to say that we'd been pressured into signing the petition and we would now like our names removed.

It was about this time that we both started to feel tired and unwell. For a while we had to ease up on going to church all the time. We thought it was best to spend a bit more time at home, reading and resting.

⁓

At work I started noticing, and eventually talking to, one of the trainee meteorologists. He had dreamy eyes and a nice voice, and he dressed like someone from a Dunedin music video. When we finally got up the courage to sit with him at lunchtime, we discovered he was smart and silly and his name was Warren. He told us he played soccer and bass guitar, and best of all he had a Mark 1 Cortina, which was my favourite car. Then one of the other trainees joined us, and when we introduced ourselves and said we'd been friends since we were two, Peter said, 'What? I thought you were sisters!'

'Oh no, everyone used to think that – especially when we had hair about this long.' I indicated with my fingers.

'What? How long was your hair?' asked Warren.

'Just a couple of centimetres. That was during our punk phase.'

'Oh, I see. What phase are you in now?'

'Oh, we're Christians now, but it's not a phase this time.'

'Christians? Not *born-again* Christians?'

We sat for two hours talking about *everything* – Christianity, opera singing, nuclear winter – and invited them to come with us to the Barry Smith crusade that evening. They said they would, but we went back to work without making a plan. To help God make it happen, I went back and forth to the photocopier outside their classroom all afternoon, and the third time I went down Warren was in the corridor. Talking about the weather is okay when that's what everyone you work with thinks about all day, so I complained to him that it was raining and we were going to get wet heading down to the crusade after work.

'Oh yeah, we're going too,' he answered. 'Do you want a ride?'

Everything was going to plan. I said yes without the slightest hesitation, then got even bolder and said we'd probably go and get something to eat first.

That's how we ended up all going to dinner at the Mexican Cantina before the crusade, eating enchiladas and beans and spicy rice and talking over the light of a candle stuck in a Mateus rosé bottle. Warren and Peter kept asking us if we were serious about the Christian thing, trying to get us to admit we were having them on. It just made us laugh, and the more we laughed the less they believed us.

At the crusade Barry Smith talked about the coming collapse of the global monetary system, the coming one-world government, the one-world religious system. A middle-aged man in glasses and a suit, he had the quiet self-assurance of a doctor, teacher or public servant. He wasn't exactly charismatic – he'd delivered this same lecture dozens of times before, and it showed – but he got us with his confidence, his certainty, his layers and layers of evidence for the 'Satanic Luciferian Plans' for the takeover of the whole world. His research into freemasonry, witchcraft, politics, and more, had revealed to him that the Antichrist was already living among us, he said. The good news, though, was that after the Antichrist rose to power, he would be defeated by the Second Coming of Jesus. During the Rapture, all the true believers, those with pure hearts – alive and dead – would rise and meet Christ in the air. We wanted to meet Jesus, but we felt anxious about the Rapture because all the evangelists seemed to have different ideas about when it would come. Would it be before or after Jesus returned? Before or after the rise of the Antichrist? Before or after the seven years of tribulation prophesised in the Book of Revelation? We wanted to know if there would be time for our mothers and families to see the signs, to turn to God and be saved before it was too late.

All evening, as we sang songs and praised God, Warren and Peter kept turning to look at us instead of at the stage, which made us feel weird and self-conscious. For a while we mouthed the words to the songs instead of singing them. When some of our Christian friends came up to say hi at the end of the night, Warren and Peter went home.

The thing we remembered most from the crusade was Barry Smith saying that now was the time to prepare for the End Times. For that, we had to make sure that our hearts and actions were pure. We also had to have a reliable car with a full tank of gas. We didn't have our own cars, but we had bicycles and backpacks and sleeping bags and boots.

To prepare for the End Times we started reading survival manuals and collecting seeds so we could grow our own food when we were on the run. We talked about where we would go. The Tararuas? The Ureweras? The West Coast? It was exciting, planning for the future, but sometimes it made us anxious and depressed. We loved our mothers and our families and did not want to leave them and did not want them to take the Mark. But they showed no signs of being born again.

We kept talking to Warren and Peter at work, because we hoped we could lead them to the Lord. And they were fun to talk to. But sometimes we wondered why the only boys who liked us were naughty backsliders or the ones who weren't even Christians yet.

Eventually, Maz and I settled on a plan for when it all went down. With a mix of excitement and dread, we agreed that we would meet, with our gear, at the wetā cave at Percy Reserve and follow the Hutt River inland from there. We had a growing sense that things were starting to happen, Before the persecution of the Christians began, said Matthew in the New Testament, there would be 'wars and rumours of wars', famines and earthquakes. All these things were happening now. And it all felt like it was getting closer when French agents bombed the *Rainbow Warrior* in Auckland.

10

Misery Hill

Following the map on my phone, we turn off Hokitika's coastal highway and head up a steep wide road, past a lighthouse, past an old cemetery and onto a flat wide plain. One of a series of uplifted river terraces, a product of the massive plate boundary earthquakes that strike every few hundred years. One of the places the mayor thinks the town should retreat to when sea-level rise starts causing trouble.

On the flat are a scattering of old wooden buildings – what looks like a church, some houses, and some dormitory accommodation – separated by vast grassy lawns, a few hedges, some agapanthus. We follow the signs to a parking area, the centre of some sort of campus, where an array of asphalt roads converge. A sloped concrete path with metal hand railing leads to a brick building that looks to be the reception. Across the carpark are two old wooden buildings with peeling white paintwork and rusty corrugated iron roofs. We exit the car, look around, lock up. There are a lot of men here, we notice, standing around the edges of the parking area. One smokes a cigarette. One holds a large black Labrador on a lead. One leans on the railing, talking to another guy, smirking, laughing, looking around. Things slow down and go quiet as we approach the building. We're being watched.

Inside, sliding glass panels separate the reception desk from the foyer. The room is decorated with the usual backpackers' paraphernalia – a string of flags, a smokefree sign, a stand of brochures advertising local tourist attractions. Less familiar is a collection of items that look like they've been made with a 3D printer – a white T-800 terminator head, a couple of skulls, a wolf mask – on a display stand. There's a handbell on the counter. We ring it and soon a man with a beard, a knee-length Swanndri, and black boots appears and introduces himself as Buck. After saying we want two beds for the night, Maz tells him her grandfather spent many years up here.

'Was he in the asylum, or the prison?' asks Buck.

'The asylum.'

We didn't realise there was a prison here too. Buck recommends a book to us, *Misery Hill*, about the Hokitika Gaol, which tells the story of the gaol from its origins in the Hokitika township to its expansion up here on Seaview Terrace. There's a copy he can lend us later. 'It's a good read,' he says.

Buck lays out our sleeping options – a room in the main lodge building we're in now, a bed in one of the old wards, an isolation cell. We're not sure, so accept his offer of a tour of the campus. The institution variously known as the Seaview Lunatic Asylum, or Westland Hospital, opened in 1872 and by the 1950s had hundreds of patients. It closed in 2009, with 22 patients remaining, but it's very mid-century in style and decrepitude. Buck shows us through a few wooden buildings, pointing out the shared dormitory-style accommodation in the ward rooms, the isolation wing with its padded cells, the rooms with plastic windows that can't be shattered and made into cutting tools. There are shared kitchens, bathrooms, lounges, a table tennis room. In the room he calls the 'drum-and-bass room' a couple of men are playing pool, while another hovers over an iPhone plugged into some huge speakers. There are high tables with stools, a couch. Around the walls are psychedelic murals – the Cheshire cat, the Mad Hatter, some skulls, and an unnerving panel crammed with ghoulish faces. Buck tells us this is where people come to hang out, have a drink, listen to music, and says we're very welcome to come over tonight.

'Cool! We'll come over for sure,' says Maz with a smile, just as I'm thinking no fucking way.

Then we're back down a corridor where Buck ushers us into a small room with a huge vertical bolt on the outside of the door, and an observation hole through which the screws, or the orderlies, or the doctors – I'm not sure if this was a gaol or asylum building – would check on the incarcerated inhabitant.

'There used to be 580 patients here,' Buck says. 'Now there are only two,' he adds, looking back at us with an impish grin. Maz laughs and I giggle nervously.

On the way back to reception we pass a series of rooms with glass-windowed walls, behind which remnants of the original hospital wards are preserved. In one diorama, a mannequin lies in a hospital bed, with what looks like electrodes attached to its head.

Maz asks if there are any ghosts up here.

'No ghosts,' Buck assures us. 'I've been here 270 years and I've seen nothing.'

As we head back to reception we thank him for the entertaining tour and exclaim over what a remarkable place it is.

'Mate, you want to see the other buildings,' he says. 'Ten o'clock tonight, hahaha, I'll take you on a bit of a tour.'

⁓

Maz is excited about sleeping in the spooky patients' quarters but I insist on a door I can lock from the inside, so she agrees to stay with me in the 'lodge', the old hospital administration block, where Buck, his partner Billie, and a few others stay. We ask about a five-bed corner room, between the shared lounge and the kitchen.

'I'll give it to you for $80. It's a nice space.'

We're sold.

The long rectangular room holds five single beds, two wooden dressers and a small round table and chairs. There are framed prints on the wall – landscapes and floral arrangements – windows that look south over the grassy terrace and west towards the ocean, and a 1960s-era patterned carpet – shades of grey, white, brown, yellow. There's a free-standing oil heater ready to keep us warm through the winter night.

We carry our stuff in from the car then I drive down the hill with instructions to buy a bottle of pinot noir – 'Martinborough, not Otago or Marlborough,' says Maz – and something for dinner.

When I return, incense is burning in the hallway and the lodge smells like jasmine or patchouli, scents that remind me of my teens. Maz is in the lounge by the window overlooking the sea, reading *Misery Hill* and playing her harmonica.

'This was the only Martinborough pinot noir,' I say, offering a bottle of Russian Jack I found at the supermarket for $23.99.

'That's good shit,' she says, taking her lips off the harp for a moment. Then she's back to it. She's not playing a tune, it's more like scales – she's trying to blow through one hole at a time – but it's okay. It's mournful, tuneless, effortful without achieving much, but it feels appropriate. I can imagine some of the inmates playing a harmonica. Is that the right word?

What were they? Inmates? Prisoners? Patients?

I grab some glasses and plates from the kitchen, then we sip our wine and eat stuffed baby peppers while the sun sets over the Tasman Sea. Maz tells me there are some deep canyons off the coast, and when the Alpine Fault next ruptures it will set off submarine landslides that could send floods back towards the town.

While we're musing about submarine landslides and tsunamis – we agree this would be a prime viewing spot if a tsunami rolled in – a tall man, a bit younger than us, enters the room. He introduces himself as Aidan, the manager, and once he can see that we're open to chatting, pulls up a chair. He is dressed for outdoor work, a parka with hi-vis strips, a cap, jeans, and lace-up boots. He's followed in by a small dog he calls Zeus, a Jack Russell poodle cross. We chat for a bit, tell him what we've been up to today, about the tour with Buck. He asks if we liked the ECT diorama.

'It's actually a defibrillator,' he says.

We laugh, but I wonder what they'd do if someone had a heart attack and needed the defibrillator.

This wasn't just an insane asylum and a prison, he tells us. It was a full-blown hospital with maternity wards, dementia wing, the works. This hostel is not for everyone – some people get scared; they don't like it.

'I guess it's not like a normal hotel where you can lock yourself in your room with your TV and en suite,' says Maz.

'No, it's more like the Hotel California,' says Aidan and lets out a raucous laugh. 'Some people are really scared of the spirits,' he says when he's settled down. A lot of people died here – in the hospital, and the prison. And there were injustices. 'Te Whiti's gang were here, illegally incarcerated by the Crown.'

He's referring to when, in the 1870s, the government started surveying Māori land in the Taranaki region, to sell to European settlers and under Te Whiti's leadership, the men of Parihaka began pulling up survey pegs and ploughing the land. Hundreds were arrested, some of them were shipped to the South Island, and many of them ended up here, in the Hokitika Gaol, where they cleared bush and widened the road up to the Seaview Terrace.

We ask about the people we encountered in the carpark and in the drum-and-bass room.

'They live here,' says Aidan. While pre-COVID he welcomed backpackers from all over the world, today the tourists have dried up and he's opened the place to permanent residents. People working nearby and earning minimum wage, or unemployed, or on a sickness benefit, looking for an affordable place to live. There are about 40 permanents here now. 'It's working out pretty well,' he says, but he's looking forward to the return of the tourists. 'That's the fun part, meeting people from all around the world, telling them stories, taking them on tours. People love the place, they come for a night and stay for a week.'

Maz and Aidan start talking West Coast stuff, local characters, the best place for a pie, a bit of engineering shit. I'm mostly quiet, but I interject to tell Aidan that I'm a writer, to explain why I'm writing in my notebook.

'That's fine. Write whatever you want,' he says. Then he says something that makes me lift my head from the page.

'The last writer who stayed here dropped dead on me. Right there, in that kitchen,' he says, nodding towards the room where I have just deposited our groceries. 'Brain tumours. She'd already written some plays and books. She loved this place and was starting to write a story about it.'

We're nodding, attentive and wide-eyed.

'This is the room where she did her writing,' he says. I look around the room at the aubergine walls, the picture windows that look past tī kōuka and harakeke to the ocean, a choice of straight-backed chair at a table or a comfortable chair, a cabinet of mismatched glassware and crockery from the mid-twentieth century. It would be a nice writing room.

'She was planning to go back to the UK to take up her writing again. It was quite fascinating.'

'Her dying?' I ask.

'No, her life. Her dying wasn't fascinating.'

Aidan had been outside mowing the lawns. When he came inside to start putting his dinner on, he found her lying on the kitchen floor. She was dead, but still warm. Then the reception bell went, so Aidan walked down the hall to see who was checking in. Seeing the look on Aidan's face, the prospective guest asked if he was okay, if he needed help with something. 'Yeah I do actually, I've got a dead woman in my kitchen,' he said. They rang an ambulance, who told them to start CPR.

We are transfixed by Aidan's story. As he continues, Maz offers affirmative gasps and words – 'Aaaaargh', 'Jesus' – and I find myself quietly and inappropriately laughing.

'So she had a haemorrhage or something like that?' says Maz.

'Yeah.'

'How horrible, I'm sorry you had to go through that,' she says. Then she looks over at me, calls me out. 'Bit of inappropriate laughter there, Rebecca.'

I'm relieved to have it named. I stop myself.

'Funnily enough, the guy who helped me,' says Aidan, 'he didn't stay. He buggered off.'

We pause to pay attention to our drinks. Maz and I drain our glasses of pinot noir, and Aidan finishes his can of Canadian Club & Dry – he has more cans in his pockets – then asks if we are up for a glow-worm tour. He's got a spot, something you won't find on any map, with better glow worms than any of the places you have to pay to visit.

It's a toss-up between a night tour of the asylum's abandoned buildings and Aidan's best-in-New Zealand glow-worm tour. We opt for the glow worms.

We meet Aidan at reception where he checks that we're appropriately dressed – jeans, boots, warm jackets, trapper hats, that'll do – then tells us to follow him. I'm excited about the tour.

First, we head over to Aidan's digs in another building. Inside a wooden institutional block, down a corridor, through a locked door, is his place – a living room, two bedrooms, a kitchen and a bathroom. In the living room, a gaming chair is set before a table holding three screens and a steering wheel. Maz and I have blown our survivalist credentials by not bringing torches so Aidan is looking for some we can borrow. As he searches, Zeus darts around and a grey cat – Puss Puss, who Aidan refers to as *beautiful boy* in a high-pitched voice – slinks over the back of the couch towards us, demanding pats.

When we're all set to go, we head outside to the car, a grunty black beast. I sit in the front passenger seat, Maz sits in the back, Zeus on Aidan's knee. A desiccated sparrow sits on the front dashboard as either decoration or talisman.

As we drive away from the building, we realise how dark it is, the only light coming from the car's headlights. Then Aidan says, 'I'm taking a bit of a detour,' and drives us to a house with a faint glow behind closed curtains and in sight of a small bonfire. I feel uneasy – what are we doing here? – but we go with it, get out of the car. Aidan wanders off and Maz and I look at each other, shrug and walk over to the bonfire. We're pleased to find Buck there, feeding sticks into the blaze. There are stars in the sky. We get talking to Buck's partner Billie, who is warm and welcoming and is wearing the same hi-vis parka as Aidan. Maz tells Billie about her grandfather, and the years he spent years living in the asylum, rather than at home with his family. An iodine deficiency, caused by New Zealand's depleted soils, had caused his mental health problems, as well as a goitre. Once the doctors identified the problem it was a quick fix. I talk about my grandmother Trudy, who grew up in the town below, and her mother Lydia, the daughter of a barmaid and a miner – Marian and Edward – who each sailed here from Australia in the 1860s. They're buried, along with the three of their nine children who didn't reach adulthood, in the Greymouth cemetery.

'You're coming home, you've got bones here,' she says.

It's generous of her to say so. My bones here go back four generations. Her ancestors arrived on waka hundreds of years ago. We talk, uninhibited by the dark, inspired by the firelight, about where our ancestors' bones are buried, about family trauma. She asks if we've visited the urupā, and starts talking about Te Kooti and Te Whiti, the massacre at Parihaka, the violence our Māori and European ancestors warred upon each other. We talk about the energy of this place. It's not for everyone.

'This is not a place for people who are unstable, or who have addictions,' says Billie and we understand. We decided when we arrived we would keep things low key. A bottle of wine to share. Not a place for a sneaky joint. Partly out of respect for the history of this place, but also you wouldn't want to feel paranoid up here, wouldn't want to start thinking about spirits, visitations, possessions.

Aidan seems to have finished whatever he came here to do and walks up to us.

'We might as well go see those glow worms.'

We get back into the car, then one of the back doors opens and a big guy gets in, moving a pizza box to make space. He is much younger than us, wearing a work fleece, shorts and boots. He carries three bottles of Corona.

'Jason,' he says, and offers Maz a beer. She hesitates, then politely declines – she won't deprive him of one of his beers. He puts one bottle in each pocket of his fleece and starts drinking out of the third.

As we pull out onto the road, I turn to smile at Maz, raise my eyebrows to see if she's okay. She smiles back. We're feeling good, about to have an adventure. Something about what we're doing, driving off into the dark with some boys, on some escapade that's possibly stretching the bounds of safety and legality, feels *very* familiar.

We drive at a slow pace across the dark plateau. There is no moon tonight. There are just the stars above us, distant glows from inhabited buildings on the terrace, and the headlights of the car. I guess Aidan knows the landscape, knows where he's going, but we might as well be flying through space.

'There they are,' says Aidan, as we pass a dark bush sparkling with lights.

'Oh!' Maz and I gasp. Then we realise it's a string of fairy lights woven through someone's hedge and Aidan is having us on.

We continue on to a place that seems like no place. When we get out of the car, it is dark all around us, and the only sense of separation between ground and sky comes from the feeling of my boots on the ground. There are a few lights in the distance, but not much else, nothing to give any sense of perspective. We turn on our torches and follow Aidan across some long, wet grass, and into the void. I stop to retie a boot lace while, one by one, Aidan, then Maz, then Jason, disappear. I can hear their voices, but the lights of their torches have gone. I hurry to catch up, then follow them down a steep bank, along a rough path through the bush, holding a torch in one hand and grabbing branches with the other.

We pass some twisted tubes of metal poking up from the dirt and Maz wonders out loud what they're doing down here.

'It's an old rubbish dump', says Aidan, telling us a bit about what he's found down here – an old filing cabinet, a few functional chairs, lots of old china. I'm starting to wonder about the health and safety implications of our adventure. Through the early twentieth century, all sorts of shit used to get dumped over banks, into rivers, into the ocean, with no thought for its environmental or health impacts. Asbestos, toxic chemicals, heavy metals, chuck it in. What else might a prison and a psychiatric hospital put in their illegal rubbish dump? Not for the first time, I wonder if, in our quest for a bit of fun and adventure, cutting loose, *fucking enjoying ourselves*, Maz and I are failing to exercise any sort of caution at all.

Then the path turns into a scramble, and we take turns holding each other's torches while one by one – Aidan, Maz, then me – we abseil down a short cliff, holding onto a rope hanging from a tree growing perpendicular to the slope. Jason, Corona in one hand, comes last. I compliment him on his ability to hold a beer while abseiling, but this is nothing, he says. 'I've been down slopes like this with a chainsaw still running.' I believe him, and I'm very impressed.

We stop at the bottom of the bank to look around. A small creek with a brown stony bottom fills the gully floor. Zeus, who was first down the bank, darts around like a little sprite, up the stream, down the stream, around our legs and back upstream. I guess he wants us to follow him.

We start walking upstream, stepping carefully on rocks to avoid immersing our boots in the water and dodging low-hanging branches of māhoe and ponga. When Zeus starts barking and running in circles, I ask what he's found.

'It's the freshwater cray, he thinks he can catch them, *don't you boy?*' says Aidan. 'Hold on,' he says, and bends down and fishes around in the water. He lifts his arm, his sleeve dripping wet, and holds his hand out to us. Maz shines her torch on a crawlie. It's like a tiny crayfish, maybe ten centimetres long, a murky greeny-brown colour. It's squirming in Aidan's grip.

'Put it back!' I say, feeling concerned for the crawlie. Aidan slides it back into the river, then points to a movement in the water.

'A cave wētā,' he says. 'They can trap a bubble of air and take it with them underwater.'

'Sounds like you've been doing some research,' says Jason. I wonder about his inflection of the word *research*, like it's a bad thing, something to be embarrassed or ashamed of. I'm pleased I'd introduced myself as a writer, rather than a researcher or academic.

'Not research, *observation*,' says Aidan. 'I've seen it.'

I hold back on saying that observation, rather than reading things on the internet, is the basis of most scientific research. Now doesn't seem the time for a philosophical discussion about the nature of science. It's dark and it's cold, and I've no idea where the fuck we are.

While Jason stops to roll a durry, Maz and I carry on up the stream, following the torchlight and the sweet smell of Aidan's caramel-flavoured vape. Maz, just ahead of me, flicks past a branch that swings back and hits me across the bridge of my nose.

'Ow! Fuck.'

'Shit,' she says. 'Sorry.'

There's a bit of blood but I'm okay. I hold my hand over my nose and look around. On the banks either side of us are remnants of the old dump. An institutional chair. A bottle perched on a metal lawnmower. God knows what's buried beneath the dirt and leaf litter.

Jason, who has caught up with us, starts talking about overburden, rounded rocks, gold mining. He now works as a moss cutter, gathering sphagnum moss, but did a stint with an alluvial gold mining operation. He seems to know what he's talking about.

'Maz used to be a mining engineer!' I offer, wondering if she's going to join the conversation, but no one responds. Perhaps they don't believe me.

Where are we going? We continue up the stony stream, but now the cliffs either side of us are vertical, covered with dripping ferns and mosses. At the top of the cliffs, maybe four metres high, trees reach out over the stream, covering most of the sky. Through the gaps we can see stars.

We're commenting on what an amazing place this is, and asking Aidan how he found it, but we're thinking, When will this ever end? Are we just walking or are we going somewhere? A subterranean cavern? A padded cell?

'I brought a single girl down here once, a European backpacker – that was a high-trust situation,' says Aidan. 'That's why I have a lot of backpacks,' he says and starts laughing at what we hope is a joke. Maz and I are close to each other now, and not for the first time I think about making a run for it in the dark. But I don't want to overreact, I don't want to offend anyone by letting on that we don't feel particularly safe, for even contemplating they might have nefarious intentions.

'Fark,' I say, just loud enough for Maz to hear.

'Hey, enough of that,' she calls out to Aidan.

'Yeah,' chimes in Jason, who is now walking ahead of me. He holds a branch back for me and I decide that staying close to him might be our best chance of survival. He seems capable and trustworthy. Chainsaw in one hand, beer in the other.

'We're here,' says Aidan, finally. We're at the end of the stream. Ahead of us is an ancient looking stone wall, where water falls from a pipe and pools in a pond just a few metres across. We turn off our torches and on the moss-covered vertical walls around us thousands of pinpricks of light emerge, constellations of stars.

We are surrounded by glow worms, more than I've ever seen – *ten times* more than I've ever seen – like we're floating through space. We stand, heads tipped back, turning in slow circles, exclaiming, while Aidan vapes.

I fill my lungs with the earthy damp air, the chuckle of the water and the bright white stars against the black night transporting me into a world where I'm at once deeply grounded and floating in space. I feel a connection with the river, the people I'm with, the planet I'm on and give thanks for this moment, this place. I'm not sure who I'm talking to, but I think I'm praying. I feel a deep sense of peace and note to myself: I want to hold on to this feeling. That's why I do this kind of shit, I remind myself. It's almost always worth following a trail, saying yes to an adventure, for what you might find at the end, and people are mostly good and kind and fun. I'm starting to miss this feeling even before it's gone.

Jason turns on his torch and we gather close to the cliff to look at the sticky silk threads hanging from the walls. The glow worms attract insect prey with their lights, then trap them in their snares. When an insect lands on them, they pull the snare up and eat the insect. They thrive in caves, grottos and deep gullies protected from the wind, where their snares can hang without getting blown around and tangled.

'You know they're not actually worms, don't you?' Jason says. We do know, but it feels less special when you know they're maggots, the larvae of a fungus gnat. The Latin name is *Arachnocampa luminosa*, the Māori name is titiwai, meaning lights reflected in water.

When I get up the next morning there's a young woman in a dressing gown sitting out a side door drinking coffee and smoking a cigarette. One of the permanents. Maz has a text from Aidan. He's remembered the name of the writer who died here so we look at her Wikipedia page. She *emerged as part of a new wave of young New Zealand writers in the 1990s*, was *a novelist, playwright and poet*. Laura Solomon died in February 2019, aged 44.

While we're reading about her, Billie arrives.

'Morning darling,' she says. We tell her we're reading about Laura, the writer who died here. Billie never met her, having arrived more recently, but tells us she has felt Laura, up and down the hallways in the lodge, and performed blessings here for her all last week. She wants to do more. She's talking about getting a group of people to do the whole place, 'a Māori blessing', not just because of Laura, but because of the whole history. The brutal psychiatric treatments. The deaths in prison. The rumoured unmarked graves.

We take photos together and hug before we leave.

As we head further south down the coastal highway, we talk about how Aidan and Jason feel like the sort of people we used to hang with when we were teenagers. Naughty. Unpretentious. Adventurous. Practical. With a huge sense of fun and a disregard for authority that is incredibly appealing. We were raised to be like that. Growing up in the era of Muldoon, Thatcher and Reagan, it was essential to rebel. There were serious things worth fighting against. The trouble was, at some point we got confused about what to rebel against, and found ourselves rebelling against our progressive, secular, liberal upbringings. It was a couple of years before we came to our senses. But damage had been done.

11

Testing my faith

Six months into my job, I moved to Christchurch for three months, to train as a meteorological observer at the Aviation College. I learned how to launch a weather balloon, make a meteorological observation, and touch type. I lived in a big house with my father, stepmother, step siblings, and the new baby. I slept in a long narrow room off the kitchen, just wide enough for a wire spring bed beneath the sash window. Outside there were chickens in a coop, firewood stacked in the shed, and winter vegetables and herbs in the garden. My mother said it would be good for me to spend time with my father. But away from my friends, away from my *we*, I began to unravel.

The first church I tried was the New Life Centre, where the services were held in an enormous auditorium with tiered seating. It was crowded and loud and everyone smiled, but I went there for three Sundays in a row before anyone spoke to me. When someone did notice me, I was invited to a Tuesday night meeting, then a Friday night potluck dinner. Everyone was nice, but with my chin-length bob, dark clothes and oversized vintage coat – Wellington style – I stood out amongst the fluffy blow-waves, pastel separates and fob chains. I felt more alone than when I was by myself.

Then I went to a service at the Assembly of God, where about 150 people had gathered. Whenever I was alone for a second someone would grab me, introduce themselves, introduce everyone around them. It was too much. It made me feel anxious and panicky and miss my friend Maz, who could talk enough for the both of us, give me space to be me. Someone asked me to fill out a Welcome Card with my name and address. Two days later I got a phone call from a Home Fellowship leader, inviting me to join them for weekly meetings. But I felt stifled, and irritated, as if my privacy had been invaded.

Whatever church I went to, the Christchurch Christians were outwardly friendly but their words sounded hollow and clichéd. I didn't know if I wanted new friends or wanted to be left alone. I didn't know if it was a Christchurch thing, or a me thing.

Because I couldn't find any Christchurch Christians to connect with, I spent a lot of time at home, in my room, reading or studying. I learned the names of different cloud formations, when and why cyclones formed, the atmospheric conditions that precipitated snow. A girl in my class lent me a Clive Cussler book and soon I was getting through one a week, borrowing them from the library, books with titles like *Pacific Vortex!*, *Deep Six* and *Cyclops* about the adventures of Dirk Pitt, death-defying adventurer and deep-sea expert. I also started reading C. S. Lewis. I'd loved his Narnia series as a child, without any notion of them as Christian parable, and now I read his nonfiction, working my way through *Mere Christianity*, *The Pilgrim's Regress* and, best of all, his autobiography *Surprised by Joy*. Inspired by his love of literature, my weekly library hauls diversified, and I started coming home armed with serious books in which it seemed everyone was some kind of Christian: nineteenth-century Russian novels, a biography of the Brontës, and *The Works of Samuel Johnson*. I wanted to know everything about everything. I held on to my pastor's advice that 'everything good comes from God' in exploring secular things and made occasional weekend visits to museums, art galleries and gardens. I prayed about what I should study at university. History and literature? Or science? After months of studying to be a meteorological technician and loving the theory but being a bit shit at the practical work, I decided that nature, the planet, the universe – God's creation – was what I was most interested in and wanted to study. There were books I could read if I wanted to know more about literature and history.

One day the younger of the two men who ran our course left the classroom to take a phone call in the office. Suddenly he was shouting, calling out numbers over the phone. My classmates and I were taking a test – in rows of desks, heads bowed over test papers – but we stopped working, wondering what was happening. A few minutes later our teacher burst into the classroom, fists held high over his head.

'I've just won a million bucks!' he shouted, then left the building.

Three days later he pulled up outside our classroom in a brand-new Holden Commodore. After the morning's lessons, he took the whole class out for lunch at the local pub to celebrate his Golden Kiwi win. We were allowed anything on the menu. Steak and chips, fish and chips, chicken and chips, schnitzel and chips. I chose the battered oysters and chips. The white-haired man who was our senior teacher ordered 'just the pub meal' and seemed quite happy with his lamb chops.

I wasn't happy, though. Why would he choose chops if

he could have oysters? Or steak? It bugged me. Why would he get the same thing he always had, rather than take this opportunity to expand his horizons?

As we ate, everyone talked about what they would do if they won a million bucks. I listened but didn't join in. I knew my problems couldn't be fixed with money.

I had begun to feel like I was existing just outside of my body, a step to the right and slightly off the ground, observing, connected to my body but not fully participating in what it was doing. I wondered if I'd left the Holy Spirit behind in Wellington. I was attentive and involved in my classes, but at home in the evenings and weekends I mostly stopped talking. There was too much going on in my head for me to choose the words. My father, after ascertaining that I was not 'on drugs', told me about being brought up in the Methodist church and said he thought a lot about religion, and philosophy, and what happened when you died. But there were no easy answers, he said, and he didn't think he'd make up his mind about it all until he was a very old man. He said he was concerned about me.

If I wasn't reading or studying I was usually praying, yielding to God, asking Him to guide me, asking Him where He was. But it felt like no one was there.

Either something was wrong with *me*, and I'd grown dead to God, or none of it was real. I didn't know which was worse. I couldn't walk away from Christianity, because I couldn't rubbish everything I'd felt, seen, experienced in the last two years. But there was no point in being a dead Christian: carrying on with all the motions without being moved by the Spirit to do them.

Where is God? I wrote in my journal. *He must be here with me – His word says He is. So what am I doing wrong? And why is He taking me so near breaking point?* My spirit felt dead. It was more serious than I could express. *I mean – it's my entire life. It's eternity. I don't want to think about it because I'll start to feel hysterical.*

I wondered whether if I held on until I got back to Wellington I would be okay. I wondered if all my friends in Wellington were still praying for me. I wondered why, if He was real, God would put me through this?

About once a week Maz and Warren – the trainee meteorologist who had become my friend – and my mother would each ring me from Wellington. We'd talk on the phone for a while, but hearing their voices just made me feel more alone and more sad.

One of the phrases that people at church said when someone was having a hard time, or when something bad had happened, was 'God works in mysterious ways'. That's the phrase that came into my head when I got a letter from Warren in which he quoted Hebrew 13:5: *I will not give you up or desert you.*

It was even better in my Bible's translation:

Never will I leave you
Never will I forsake you

It made me feel warm inside, and it made me laugh. Why did God use Warren, a non-believer, to speak to me from scripture? Was it because I had such issues with other Christians, at least the Christchurch ones? It reminded me again that *all Good comes from God*, and that people could be good, even if they weren't born-again Christians.

⁓

After reading the Bible, reading C. S. Lewis, and doing a lot of thinking, I decided I'd been trying to change myself into other people's ideas of what a Christian was, rather than listening to God, following my heart, using my brain. I decided some of the Christians from our youth group were judgemental and conformist, and that when Maz and I first joined the church they didn't fully accept us because we looked – and maybe acted – different. We weren't girly, we weren't submissive, and we questioned things – several people had told me I had a 'rebellious spirit'. And then I thought about how Jesus actively sought out the people who were different, the people on the margins, the people who were unloved. And why were the Christians so hung up about homosexuality, I wondered? If they were going to obsess about something written in the Bible why not take a stand against adultery, or idolatry, women wearing hats in church, or eating the meat of cloven hooved animals? Who got to decide which bits of the Bible were to be taken literally, and which bits were just historical and cultural remnants of the times in which they were written? I read Romans 14 over and over again. It was mostly about food, but it seemed to acknowledge that we all had different ways of serving God, different ways of being *good* – a product of our upbringing and culture – and we should be true to them. *Nothing is unclean in itself. But if anyone regards something as unclean, then for that person it is unclean* it said. And, most importantly, *Therefore let us stop passing judgment on one another.*

Some of the Christians I knew seemed so sure of themselves, so sure that it was right for a woman to obey her husband, so sure it was right to oppose homosexual law reform, so sure that people who weren't saved were going to Hell. But there were problems with the logic. If all Good comes from God, and all Love comes from God, I reasoned, how could it be that God would condemn my mother and sister to hell for not being saved, while I would never do anything to hurt them? If all Good came from God, how could *I* be kinder or more compassionate than God? It didn't make sense.

I tried to take stock of the things that *I* felt were wrong, the products of my upbringing and culture and what I hoped was God – rather than religion – in my heart. As well as standard Biblical and legal things – stealing, killing and so on – the things that gave me a deep sense of unease, of *wrongness*, were the nuclear arm's race, war, racism, sexism. There were little things that upset me too. I felt an almost visceral antipathy to lying. I couldn't do it and it felt abominable when other people did it. I was grossed out by food additives, native forest destruction, processed foods, the Christchurch smog I had to ride through on my bike. And I got really hung up on something that I'd learned at a Barry Smith crusade – the government now controlled all the seeds, he had said, to avoid unwanted hybridisation. You could no longer buy a packet of seeds that would produce crops year in, year out, rather they must be bought seasonally, he said. This interference with nature's self-continuation processes made my head spin. Through human history, gardeners had been saving seeds, experimenting with seeds, sharing seeds. Was this now going to be taken away from us?

Thinking about all these things together gave me a mounting sense of hysteria, a need to escape from society. I wanted to be free to go and live somewhere remotely, away from everyone else, and grow my own vegetables. But the government was controlling the seeds and wanted to keep track of me. If I disappeared into the bush I'd have government employees tramping in after me every five years to fill out a census form.

We weren't really free at all. How can one person presume any authority over any other person? After two years as a Christian, and 13 years at school, I'd been suppressing something, and now I felt like I was going to explode. I was exhausted from the mask I'd been wearing. I began reading about anarchism and feeling stifled by anyone who claimed to have authority over me, whether parents, church, or government.

But then I got a bad flu and couldn't go to class. I went to the doctor. My stepmother brought me mugs of what she told me was 'magic koromiko tea', but I still felt nauseous and floppy after a week. My mother sent me effervescent vitamin B tablets and told me she was worried about me. After the flu I took a lot of time to get well. I was tired and listless and unmotivated. Then my dad and my stepmother and the baby left. They went on a long trip to China. And something happened to me. I lost my shit. I couldn't get out of bed and I couldn't stop crying. My mother rang her Christchurch friends, and one of them drove me to my old aunty's house, where she brought me breakfast in bed in the mornings – thickly sliced white bread, toasted with lots of butter and Vegemite – and we ate soup and casseroles in front of TV in the evenings. In the weekend she gave me craft projects to do, and sat me down with coloured paper and felt pens to make pictures for the wall of the library where she worked.

My mother asked me if I needed her to fly to Christchurch, but I knew she had to work, and couldn't really afford it, and if I held on just a bit longer I'd be on my way home.

I didn't think and I didn't read and I didn't do much of anything and after two weeks there I couldn't even remember what had happened. My mum contacted my dad and they decided that maybe I should see a counsellor. It had been five years since I'd had family therapy, as part of my Wellington blended family, and I agreed it might be good for me to have some sessions on my own.

12

Rupture

A week into our journey, and we've regressed. While I made an effort to scrub up for my interview, we're going days between showers and having coffee, pies and lollies for breakfast. We no longer speak in sentences, satisfied with saying a word or two to get the desired effect on the other, usually laughter. We're drinking at inappropriate times. We keep losing our stash. I've taken to carrying my go bag around. I'm still reading *Unsheltered*, and I identify with the heroine, searching for her child in a climate-ravaged future world, focused on survival with all her stuff in a backpack.

'There's so much gold underneath this fucken town,' Maz says as we drive into Ross.

We stop at the Ross Museum. *I struck it lucky in ROSS GOLDFIELDS NZ* says a painted display of a gold miner. *We had a ball in ROSS GOLDFIELDS NZ*, it says above a row of dancers doing the cancan. It's a common presentation of European settlers to these parts – miners and merchants, dancers and barmaids. A bit like *The Luminaries*, and pretty much like my ancestors, who arrived in the 1860s. On the 1871 marriage certificate of my great-great-grandparents Edward Garland and Marian Milne, his occupation is given

185

as 'miner' and hers is 'barmaid'. Edward was 27, Marian 23. The family story is that she worked in hotels making beds, so perhaps she progressed from bedmaking to tending bar. I try to imagine being a West Coast barmaid in 1871, with no other family in New Zealand. Wild.

The Ross Museum tells us that from 1865 to 1914 the Ross goldfield was the most productive in New Zealand. The first gold, found in 1864 in 'the rich gravels of Jones Creek', was recovered manually, after which 'shaft mines were established to work deep deposits in the flats and a giant sluicing operation attacked the terraces'. By 1865 Ross had a population of 3000, living in tents or wooden houses, and a township with churches, bars, a gaol and a hospital was well on the way to being established.

As the gold rushes attracted miners and other settlers to Otago and Westland, gold export duties and licences were used to fund public works – roads, ports, civic buildings. One caption in the museum notes that *Large tracts of the environment were permanently destroyed – damaged by mining techniques*. When my ancestors arrived in New Zealand, they sailed to Lyttelton then made their way to the West Coast. Edward had come to try his luck on the goldfields, and Marian because she had no choice. The family story is that her mother died, and after her father remarried, her young stepmother put her on a boat from Tasmania to New Zealand, telling her she was on her way home to England. She was 16 years old.

Maz is right about the gold. The town sits above a goldfield that in 1993 was estimated to be worth $700 million. An aerial photograph shows the mine workings, just east of the town where, between 1989 and 2004, the gravels were reworked using open cast methods to extract every last remnant of gold. In the photo it's a giant pit, about the size of the town, surrounded by lush green bush.

As we continue south, the road takes us inland, close to where the West Coast river flats meet the mountains, past lakes, over gravelly rivers, through small towns.

At Whataroa, 'Gateway to South Westland', Maz checks out the local café for supplies, while I look in a shop window where there's a sign: *Alpine Fault exposed here.* The shop looks to have been closed for some time. Inside there is a 3D model of the Pacific Plate colliding with the Australian Plate, the Alpine Fault an angled plane between the two. The shop promises tours up the nearby Waitangitāhuna Valley, to Gaunt Creek, to see the exposed fault – a sight 'like nothing you've seen before'.

The Alpine Fault is visible from space. It's the biggest faultline in New Zealand, the sharply delineated western edge of the snow-covered Southern Alps. It was discovered in 1941, by geologists Harold Wellman and Dick Willett, before the science of plate tectonics was even established. Then, in 1948, inspired by a geological map of New Zealand where the South Island rocks were bisected by this massive faultline, Wellman proposed the then radical idea that the Alpine Fault had moved 480 kilometres, laterally, over millions of years, dividing rocks that were once together.

In the window is a poster presenting research by some geologists whose names I know, telling the story of the fault and its impact. The most recent fault rupture was in 1717. It typically moves with big quakes, 7.3 to 8.3 on the Richter scale, with horizontal movement of eight metres, vertical movement of two to three metres, and ruptures along 400 kilometres of the faultline. Given the size of the quakes, and the high rainfall, the posters say that 'extremely large volumes of sediment could be released in large earthquake events and could be rapidly delivered to the coastal lowlands'. Rapid delivery means they'd flow down waterways. You'd want to stay away, not go stand on a bridge – if any were left intact – to take a look. You'd obviously want to be clear of the faultline, plus any steep slopes that looked like they could fail. Maybe the coast too, in case landslides in submarine canyons led to a tsunami.

I make a mental note of the hazards, which I was already broadly aware of, then return to the car.

Maz eats a pepper steak pie while we continue on down the Glacier Highway.

'Fuck yeah,' she says as she screws up the paper packet and sauce sachet and chucks them onto the floor next to her feet.

The West Coast highway, State Highway 6, goes as far south as Haast, after which it's a choice of heading east over Haast Pass to Wānaka or continuing south to Jackson Bay, a coastal dead end before the wilds of Fiordland. We're going as far as Franz Josef, a tourist town about halfway between the coast and the Franz Josef Glacier. I want to see the melting glacier. Along with the nearby Fox Glacier, the Franz Josef has experienced massive retreat in recent years.

We pass signs advertising glacier flights, scenic helicopter tours, nature walks. As we drive over wide gravelly rivers or below steep bluffs, Maz taunts me by saying, 'Imagine if the Alpine Fault went *now*.' We assess our chances of survival if the bridge collapses, how long it will take for rocks to tumble down the riverway, if a cliffside will fail and bury our car in rocks.

Maz drops me at the Franz Josef Visitor Centre and heads off to find us somewhere to stay. I want to know more about the glaciers. I know from a colleague's work, and from news reports, that the glaciers have experienced dramatic recent retreat, but I'm interested to hear about it from a local perspective. I was last here more than 20 years ago, on a road trip with my now-husband, Jonathan. I have photo of us at a viewing point, the glacier looming behind us. I get chatting to Carl, a local mountain guide, who knows the glaciers well.'

'What's changed since I was last here?' I ask him.

Carl primes me to be 'surprised, startled, depressed' at what's happened to the glacier. 'You would have had that intimate interaction without the aid of a helicopter, ice axe and crampons. Whereas now you can only experience it from afar.' The closest viewing point puts visitors almost three kilometres from the terminal face of the glacier. If you don't want to pay for a helicopter trip, it's the nearest you can get, he says, given the various glacial and fluvial hazards.

According to glaciologist Brian Anderson, the glacier has retreated almost two kilometres since I was last here in the late 1990s. I knew about the retreat, but didn't realise you can't walk to the new terminal face of the glacier. I ask Carl why the track and viewing platform haven't moved to follow the glacier.

It's about safety, he says. The mounds of terminal and lateral moraine are prone to collapsing, the rock walls newly exposed after the ice retreat are unstable, the terminal face of the glacier is dynamic, and the river flowing from the end of the melting glacier has a tendency to change its course, to jump from one side of the valley to the other, taking out the walking track when it moves.

A couple of years back, some walkers were stranded on a gravel island in the middle of the river after heavy rain caused it to rise quickly and burst one of its banks. There have been other accidents here and at Fox Glacier over the years. Tourists crushed by falling ice after crossing safety barriers at the terminal face of the glacier. Drownings in the Waiho River.

Most accidents come from people ignoring warning signs.

To avoid risk to tourists, the viewing platform has been moved *down* the valley at the same time as the glacier has retreated *up* the valley. Carl acknowledges that this is disappointing – and sometimes controversial. I recall what the Westland mayor said – 'They need to put a bulldozer in' and flick the track over to the other side of the valley. The problem with making a new road, says Carl, is that it would cost millions of dollars, and there would be a risk that it would be taken out again in a few years. With a limited conservation budget, and so many demands for that money, it's a tough argument to prioritise a solution that is possibly unsafe and short-lived. Similar things are happening at Fox Glacier, Carl says, where the track up is blocked by a massive landslide.

'But what does this mean for the whole region?' I ask. There are signs everywhere saying *Welcome to Glacier Country*, but the glaciers are disappearing. He agrees it's an issue. Some people in the community say it could be a golden opportunity to focus on the other gems of the West Coast, like the flora and fauna. Zero Invasive Predators, or ZIP, are working in the Perth Whataroa area with an eye on expanding out to Ōkārito, and there's a new trapping project around Lake Matheson with Project Early Bird. Government-funded predator-free programmes are providing jobs, and people are working to restore the natural biodiversity that has been ravaged by rats, possums, stoats. They're bringing back the birds, says Carl. Farmers, community groups, schools, are all getting involved.

What's happening to the glaciers is dramatic and highly visible, but what about the impact of climate change more generally? Being so close to the Waiho River, the whole town is susceptible to flooding, and there are questions about how long the rock and gravel stopbanks will last. Along the coast, there's risk of sea-level rise. But it's a very well-educated community, Carl says, with strong local understanding of climate change.

If we want to visit the glacier, we'll have a 20-minute walk from the carpark to the viewing platform. If we wanted a closer encounter, we'd need to hire a helicopter, or ignore warning signs and go it alone. Alpine guides won't take tourists up anymore because of the risks. So the only people who go onto the glacier are equipped with ice axes, crampons, and extensive experience in the alpine environment. I assure him that we're not that kind of visitor.

'So the only way to see the melting glaciers is to burn a whole lot of fossil fuels in a helicopter?' This all shrieks of last-chance tourism, I think. It's like the tourists who visit Antarctica. See the melting ice, the penguins, the whales – before it's too late!

'It's a balancing act,' Carl says. 'If nobody is going to experience the glacier, then what significance would they give to it? What understanding might people have of climate change?'

We decide to drive up to the glacier before settling into our digs. We park up and follow a track through the lowland rainforest, a cheeky little tomtit hopping ahead of us, flitting from the gravel track to a low branch, to a moss-covered bank, beckoning us on. The track takes us up Sentinel Rock, to a lookout facing east into the valley and the mountains beyond. Despite being prepped for disappointment, we are stunned by our first sighting of the glacier.

The Waiho River flows beside us in a channel that 150 years ago was filled with the glacier. In the distance, behind a forested slope, a steep tumble of bright white ice gleams in the sunlight. The glacier that used to flow as a 70-metre-deep river of ice – down to the valley floor, along the valley around corners and bends – now hangs nearly 300 metres up the cliff. It lies between deep, dark schist mountains, ancient sedimentary rocks compressed and hardened over millions of years, uplifted from deep below the surface where high temperature and pressure turned them into muscular giants. The first name given to the glacier was Kā Roimata o Hine Hukatere. The glacier is Hine's tears, frozen by the gods as a memorial to her grief after her lover was killed in an avalanche in the mountains of Westland. It seems apt. A melting glacier made from tears.

The display in front of us tells the story of the glacier's retreat, from 18,000 years ago when the glacier reached the ocean, to the first European documentation of the glacier in 1865, to now. I'm not sure when the display was made, but the final board shows a picture of the glacier high up the mountain slope. *Will this be Franz Josef Glacier in 2100?* it reads. The board needs updating, because it looks to me like that's where the Franz Josef Glacier is now, in 2021.

⌒

It's night-time, and Maz and I are sitting in a hot spa, pleasantly stoned. While Maz chats to a couple in the tub, I sit low in the water and lean my head back to let the rain drip on my face.

Afterwards, we walk back along the bush-lined path to our cabin. I start singing a White Lies song, the one about growing old together and dying at the same time. I don't know who I'm singing to – myself? Maz? – but the words feel important and I sing loudly.

After learning that the place we've chosen to stay in Franz Josef is adjacent to the Alpine Fault – 'Just up here' the host had said, pointing to a junction between the highway and a side street on a map of the small town – I've been trying to get my anxiety under control. I'm not just anxious; I'm angry at myself for getting into this situation, for not having thought about this earlier, for not choosing a safer location to stay the night. I try to engage my logical self by thinking that, statistically, one night should not be a problem. Earlier, while Maz was on her Zoom call, I drove out of town and over Stony Creek. I decided that if a quake happens we will make our way there, safe from the threat of rockfalls, flooding rivers, tsunamis. It's important to have a plan.

We eat dinner at the eco retreat restaurant – pizza served on a wooden board – by a fire roaring in a stone fireplace, but it's tiring being normal in front of other people, and we soon retreat to our cabin to get stuck into the wine. Maz connects her iPhone to the cabin's speaker and blasts out some frenetic electronic dance music she's been playing in the car. It suits this place, and we start dancing mindlessly on the porch in the dark. Gentle rain is falling around us, the only light the glowing embers at the end of the joint. Every now and then one of us laughs, partly from the fun of it and partly at how ridiculous we are, two fifty-something women reengaging with our feral selves.

It's cold, so eventually we go inside to tackle the wine and our chocolate supplies. I have another go at lighting the incense. I light the end of the stick, blow out the flame, but it refuses to take.

'Shit incense,' I say, and join Maz on the couch.

It's looking like we'll fall asleep before we finish the wine, so we pull on our pyjamas and retire to our beds. There's still a gentle rain outside. I've had a lot of wine and a fair bit of weed, but I can't sleep. Phone in hand, I start googling. I need some hard information to rely on. *New research reveals the chances of the South Island's Alpine Fault generating a damaging earthquake in the next 50 years are much higher than previously thought* says a recent report. An adrenaline surge floods my body.

The new research, published just weeks before our trip, says there's a 75 per cent chance of an Alpine Fault earthquake in the next 50 years, up from the longstanding 30 per cent figure I had in my head. I fumble on my phone in the dark, reading scientific reports and news articles and calculating the odds of an Alpine Fault movement.

Even if there is a 100 per cent chance the earthquake will happen in the next 50 years, I tell myself, that's still only one chance in 18,250 that it will happen *tonight*.

I call out to Maz to tell her the good news, but she mutters something about my calculation being flawed, about how the risk is 'not linear'.

I slump. She's right. The longer we go without a quake, the higher the chance of it happening. I can't work out if that's good news or bad, so I settle into researching the impact of an Alpine Fault event on the town of Franz Josef.

I'm excited when I find a report containing a photograph of the bulging hillside above the town, but I startle when I read that in the event of a catastrophic failure of the slope, the resultant debris *could result in a considerable portion, if not the entire town, being overrun.*

The adrenaline surges are exhausting. I fall asleep with my phone on my chest, waking periodically to the realisation that, because of the non-linear profile, the risk of an earthquake is now higher than it was when I went to sleep.

At some point in the night, I have a revelation. I call out to Maz.

'Maz, Maz . . . it's not incense. They're scent-diffuser sticks.'

I hear a muffled snort of laughter from beneath her blankets.

My scent-diffuser stick realisation is trivial, but it calms me. Maybe it's better to try to go with the flow. Sometimes we just don't have all the information, I tell myself, or we make the wrong assumptions, and we're never really fully in control.

I get back to my somnolent doomscrolling. *Parts of Franz Josef could disappear into eight-metre-wide crevasses* says one report. Aftershocks, landsides and floods will follow. If it happens at night, it will feel like Armageddon.

The next morning, I'm pleased to note there has been no earthquake, no catastrophic failure of the slope above the town, no falling rocks. I'm grumpy from lack of sleep, though, and keen to leave.

At checkout, there is a different European-accented woman at reception. I ask if the lodge has any information for guests about the earthquake hazard. She furrows her brow as if to say, 'What?'

'The Alpine Fault,' I say.

'Yes, just the same as everywhere else – drop, cover and hold.'

'Yes, but the earthquake risk here is not the same as everywhere else,' I say. 'The faultline runs through town.'

'Yes,' she says with a smile.

'So, if the earthquake happened, and we survived it, where should we go?'

'Well, I would say the place to go would be the helicopter port,' she says in her sing-song accent. 'There are many helicopters there, parked up overnight. That would be my summation, but I've been here ten years and I've only felt a few minor earthquakes.'

'You know there's a 75 per cent chance that the Alpine Fault will go in the next 50 years,' I say.

I'm determined to make my point, but I must be being irritating, because a woman behind me huffs and walks off.

'The people who live here are aware of the risk, and we're comfortable with it,' she says.

I'm astonished that people don't care. It's not just a matter of intense shaking, I want to say. Scientists have identified that an Alpine Fault event would cause a cascade of hazards, including landslides, rockfalls, liquefaction. Buildings, roads and infrastructure along the fault rupture zone would be torn apart. Hundreds of people would die, and thousands would be injured.

In the end, all I say is: 'Well, it would be really good for guests to know what to do, the safe place to go if you survive the quake.'

She agrees to pass that on.

I also tweet about my sleepless night calculating the odds of an Alpine Fault rupture. I tag in @AlpineFault8, a programme of 'scientific modelling, response planning and community engagement designed to build a collective resilience to the next Alpine Fault m8 earthquake'.

They reply: *Don't let it ruin your West Coast time, without Alpine Fault Eqs our beautiful landscape would be much less spectacular* 😎 ⛰.

I concede that it's a good point. The earthquakes are responsible for pushing up the ranges, creating the steep sides of the mountains, fjords, river gorges, waterfalls. Having survived the night, I vow to make an effort to enjoy the wild scenery in our last couple of days on the coast.

As we head out of town, I show Maz the place I drove to

the night before, my safe spot. 'I drove down here, and along here,' I say, 'and then thought, shit, I'm just getting closer to the hills, but they're lower here and there are gentler slopes.' I'm looking for reassurance, a bit of expertise. 'You know there was talk about moving the township?'

I've read that the locals didn't want to move, but even Westland's climate-change-denying mayor was adamant that the town's emergency services – the police station, the fire station, the community hall where people gather during emergencies – needed to be moved. Why the blasé attitude from the locals? All along the trip I've heard people saying they're not interested in scientific evidence, they're interested in their own observations. They haven't seen an Alpine Fault earthquake, so maybe they think they'll be okay.

'People's own observations only go back as many years as they remember, forty years or whatever,' says Maz. 'In geological time that's nothing. Even in historic time that's nothing. Our experience is so limited. I think most people don't have an understanding of risk. Which is why your science communication is really important.'

It's good of her to say so. Sometimes I'm not sure how important my work really is. Useful to some, I guess. It keeps me entertained, and paid, and gives me an excuse to have outdoor adventures. But sometimes my understanding of geological hazards or climate change sends my mind racing towards a worst-case scenario, like it did last night. While I know that these fears are more rational than fears of demon possession or an underworld lake of fire and brimstone, they can be overwhelming. I need to figure out how to acknowledge these risks, and let them inform the way I live my life, without falling into an anxiety spiral or withdrawing from the world into a life so safe that it's small.

'So I stopped just up here, what do you reckon?' We're over the river, near some houses.

'Absolutely fine!' says Maz kindly. Then she says, 'All these valleys would be subject to major liquefaction, though. They're alluvial plains.'

We discuss it for a while, then eventually settle on a spot near a small hill, on the flat but not in the middle of the alluvial plain.

I think about how, even in Wellington, away from the fault, the impact of an Alpine Fault earthquake will be significant. There will be damage to buildings and infrastructure, and disruptions to supply chains that depend on the highways, airport and ports that provide lifelines in and out of the city.

'I might get the piles checked in my house,' I say, 'make sure it's properly attached to them.'

'I read somewhere, and I don't know how true it is, that if it went off, and it was like an eight or a nine, it would be felt in Sydney. But it depends on where it starts, which direction it goes.'

Earthquakes are weird like that. The Kaikōura earthquake was felt more strongly in Wellington than in Christchurch, even though the epicentre was closer to Christchurch. An Alpine Fault rupture is most likely to occur on the stretch between Fiordland and Kaniere but could be felt more strongly in Wellington than in Christchurch.

'It's all about the angle of the dangle,' says Maz. 'Engineering term,' she adds with a smile.

We've realised that we're through. 'We're out of Dodge. It can happen now,' says Maz. It is going to happen at some stage, probably in our lifetimes. But for now we can start focusing on other things.

13

Backsliding

Despite my flu, and mental collapse, I aced my meteorological observer's course, then went home. The first thing I did was meet Maz and – after 15 weeks in the McDonald's-less South Island – go to McDonald's for a Filet-O-Fish, fries and a caramel sundae.

Without even talking about it, we started easing up on the three-times-a-week Christian stuff. Or maybe we were scared to talk about it. We were now legally old enough to drink and, as our friend Bruce pointed out, the Bible was full of people drinking wine, so we had a glass of wine when we went to dinner with some friends from church. I spent my first weekend back in Wellington at Maz's place, and on Sunday afternoon, after church, we drank glasses of Sangria from a pitcher stuffed with tangelos while her mother and friends talked and read magazines and newspapers – the local feminist magazine *Broadsheet*, the current affairs weekly *Listener* magazine, the trashy but fun *Sunday Times*.

Lying on the grass in the sunny back yard with my best friend, I felt relaxed and happy. Her mother's friends drinking with us were lesbians, a couple, people I'd known since I was a pre-schooler, and I knew my church's opposition to homosexual law reform was nonsense. These were my people.

Some of our friends from youth group had started pairing off, and we were pretty sure they weren't following our pastor's instructions to be chaste. We wondered if we'd taken everything more seriously, more literally, than others in our youth group who'd grown up with Christianity. Some of our friends backslid, stopped going to church altogether, and got right back into the bad stuff they were into before they were saved – drinking, taking drugs, getting into trouble with the police. We started making other friends – people who were smart and funny and good and wanted to talk about everything the way we did, people who didn't have a black and white view on things. Shit was happening in the world. The nuclear arms race was heating up and France continued to explode nuclear bombs, 'tests' they called them, in the Pacific. Things were still pretty fucked, but they started looking hopeful in New Zealand when a new Labour Government introduced a nuclear-free policy, stated their clear opposition to French nuclear testing, and told the American nuclear-powered warships they weren't welcome in our harbours.

I loved my job, but working as a meteorological technician had helped me figure out what I wanted to study, and I was ready to go to university. I would stay living at home in Wellington and start a degree in geography, geology, physics and maths. And a bit of Russian literature. Maz was going to university too, but in Auckland, where she would study engineering and live with her aunty. At first we didn't know how we'd cope with permanently living in different cities, but our parents said it was good. Someone said it would be a circuit-breaker.

When our former pastor, the young and kind one who used to tell us we were God's little masterpieces, visited the church one Sunday, we told him we were going to university.

'Good!' he said. 'It will teach you to think.'

We talked afterwards about the intense look, and smile, he gave us when he said that. Why would our pastor say it was a good idea to learn how to think? Wasn't university likely to open our minds and make us less likely to be Christians? Some of the other Christians from our church were at university, but they were studying things like accounting and management and we didn't think those subjects would risk teaching you to think, or open your mind, but we were going to study science and engineering and didn't expect to find many Christians in our classes.

We started going weeks without attending church.

When I did go, I felt like I was seeing the church and our new pastor with different eyes. The noisy worship and speaking in tongues seemed aggressive, competitive. I thought about how we'd been teased about our looks and our attitudes when we first came to church, and felt angry. Jesus totally hung out with the weirdos. I'd started reading books about Christianity by liberal Anglican scholars, which reminded me that Christianity was about love, and being humble, and caring for other people, and even caring for the planet we lived on. But I had doubts. Maybe it was just a cop-out on my part. Perhaps it was Satan speaking.

One day, our pastor urged us to sign another petition. It was a petition against New Zealand ratifying the United Nations Convention on the Elimination of All Forms of Discrimination Against Women. This convention would help women to have the same legal rights as men, to be paid the same as men, to be treated with the same respect as men. We knew our pastor was weird about women. We didn't even ask him to explain his stance on this one.

We knew that some Christians thought that feminism was bad. Some evangelists said its aim was domination of men and destruction of the family, and even associated it with Wicca and Satanism. But we'd read the Bible. All of it. Being a woman wasn't a sin. We lost respect for people who signed the petition because they were opposed to the ratification, and we lost respect for people who signed it mindlessly, just because they were expected to.

It was summer, and we'd mostly stopped going to church. We were hanging out with old friends – including Adam, and other friends who had led us to Christianity but had never gone to our church – spending days at the beach, listening to Bob Dylan, eating ice creams, hot chips, drinking beer.

But one Sunday we had to turn up at church, as the choir was due to lead worship. Leading worship meant that we had to sing at the front of the church, on the stage, and were meant to be visibly getting into it, inspiring others. We felt nervous, and took our friends with us for moral support.

At first, I found it easy to play the part. I sang my alto, raised my arms, prayed inside my head. But then a strange feeling came over me, and I found myself mouthing the words to the songs while staring condescendingly around the hall. Was I just caught up in some sort of mass hysteria? I had that feeling like I was attached to my body but not really in it, and it scared me.

But when Richard – who was married now – gave a talk about backsliding, I felt so bad that I went up on the altar call with Maz and other backsliders. I prayed that God would help me 'want to want Him'. I cried, but more out of confusion and self-pity than repentance.

After the altar call, as we were having coffee, older Christians came up to us, offering hugs, inviting us to come visit them for a chat, for a meal. But I felt wary, scared of being taken over, smothered, patronised.

The younger Christians, the people I had thought were my friends, looked at me in a funny way. 'Hi, Rebecca,' they said, as they walked past. I wondered if they were scared, like they'd catch something off me.

As Maz bounced around chatting to people, giving and receiving hugs, I stood back feeling sorry for myself. I so much wanted someone to talk to, to spill my guts to, but I found it hard to relate to the Christian girls, and because none of the boys felt marital towards me, I got overlooked, even as someone who wanted a friend. Maz told me afterwards that no one could cope with me looking so uncomfortable, like I was only just handling it. Apparently when I'm thinking a lot, or just feeling shy or sad, some people see me glaring or scowling.

After weeks of feeling happy, enjoying the summer, and pleased to be back in Wellington, I realised that going to church made me feel uptight and angry.

~

One night, Maz and I waited until we were a little bit drunk, then snuck into the church, giggling, and dropped our flouncy peach choir dresses, in plastic shopping bags, in the foyer. That night we got very drunk. Then we left the Church for good.

It didn't take long to find boys who liked us, who didn't think we were weird – or did, but liked us anyway. Maz moved to Auckland and got a boyfriend there and I met a boy at university in Wellington who was cute and funny and talked a lot about movies, and music, and ideas I'd never heard of, like nihilism and existentialism. They became the people we would go and see bands with, get stoned with, our new best friends.

The months and years that followed were like riding a pendulum. We put a toe back in the punk scene – we had a memorable night seeing No Tag play at Rising Sun in Auckland – but no longer felt impressed and intimidated by the older punks. I joined the Socialist Action League and the local chapter of NORML (the National Organisation for the Reform of Marijuana Laws), but was uninspired; the socialism meetings were mostly full of earnest men discussing Trotsky, and the NORML meetings were primarily an excuse to sit in a circle and smoke a joint.

I drifted away from all group activities and focused on my geology classes, which I loved. Without a true best friend to help me navigate the world, without a community wrapped tightly around me, I found myself more and more content to be alone. I was still curious, and open, and found myself attracted to many things – I tried meditation, 'past life regression', Tarot readings – but whenever I got too close and realised the fervour of the people who genuinely believed these things, it always seemed a bit ridiculous, and I backed away. My bullshit detector had become highly sensitive. I realised I had developed a strong aversion to any sort of dogma, patriarchy, groupthink, no matter who was peddling it.

＿

I have never really stopped praying. Sometimes, still, I read the Bible. I find comfort in Ecclesiastes – *everything is meaningless*! Occasionally, I've been to services in a liberal Anglican church, but never felt compelled to keep coming back. I've never been brave enough to re-enter an evangelical church. I'm not sure what it would trigger. I've had lots of therapy, including from a Catholic nun, who told me even she was uncertain about the existence of God.

I've read that for every person converted to Christianity, four leave the church. How do they decide how to live? I wonder. Do they adopt another religion? Find a secular tribe and accept their dogma? Sometimes it takes courage to be what my Methodist and Rationalist ancestors called a 'free thinker'.

One of the fields I've found myself working in, science communication, acknowledges that attempts to change people's minds about things they feel passionately about – whether it's vaccination safety, the risks of 1080, climate change – can backfire, further entrenching people's original viewpoints. Facts don't change people's minds. Can we encourage free thinking and critical thinking without giving space to toxic individualism? Can we forge ways to work together despite our uncertainties and disagreements, without the need for rigid adherence, without making our togetherness nutty and weird, without judging each other so much? Capitalism has clearly fucked things, so what's the way forward? Humanistic capitalism? Anarcho-communism? Or can we just be done with all the isms?

14

Old bones

As we continue up State Highway 6, crossing bridge after bridge, the rivers are rising.

'I'm liking this rain,' I say.

'Yeah, I would have felt cheated if we came to the Coast and didn't have any rain,' says Maz. 'Look at the debris coming down there! Jesus. Wild.' We cross a one-lane bridge over a narrow rocky stream, the steep slope a tumble of huge rocks.

Maz slows down over a windy bit of the highway. The Skypath announcement is on the radio. She's known it was coming, but has had to keep quiet about it until now. Rather than adding a cycle lane to the existing bridge, there's a new plan to build a separate walking and cycling bridge over the harbour.

'From coal miner to cycle-bridge builder. How's that for a fucken journey?'

'From wickedness to virtue,' I add.

As New Zealand's first female mining engineering graduate, Maz was the subject of my first published article – outside of student newspaper *Salient*, that is – which ran in the *NZ Listener*. I wrote about how Maz, as a female mining engineering student in the late 1980s, was unable to work underground in New Zealand, since it was illegal. She had to travel to Australia each summer to get the practical experience needed for her engineering degree.

After the Skypath announcement there is news about the floods. The road to Christchurch near Porters Pass is open again. So it looks like we will be able to head back over Arthur's Pass. 'They must have put in a Bailey bridge,' says Maz.

As we cross the Taramakau River between Kūmara Junction and Greymouth, Maz draws my attention to a two-lane road bridge beside an older dual road-and-rail bridge. 'Fulton Hogan did this one.' She admits to getting excited on seeing a new bit of engineering. I understand. I still get excited going through Arthur's Pass, on the highest road in the Southern Alps. It was built in 1999 to replace the old road that hair-pinned and zigzagged down the crumbling mountainsides and includes the Otira Viaduct – a 445-metre cantilevered bridge over the Otira River – and a stunning stretch of road topped by a sloping rock shelter, which, in heavy rain, can make it feel like you're driving behind a waterfall. Last time I crossed the pass, with my family, I got so worked up at the sight of the viaduct and the rock shelter – exhorting the children, 'Look!' – that Jonathan told me to pull over and let him drive.

We follow the train tracks inland, towards Reefton, and enter a misty plateau, filled with farmland and pine plantations.

'I can fucken smell that pie,' says Maz. As we cross the Inangahua River into Reefton, the air is thick with coal smoke mixed with cloud.

Welcome to Reefton, the town of light. Reefton was the first town in Aotearoa to be lit by electric light, powered by hydroelectricity from the Inangahua River. Mining here was different from the alluvial mining that saw men sorting through river gravels to find specks, dust, and occasional nuggets of gold. Here, miners dug shafts and tunnels. They used pickaxes to loosen lumps of gold-bearing quartz from veins running through 500-million-year-old sedimentary rocks, then they brought what they'd found to the surface to crush in a stamper battery. As she tells me about her great-great-grandfather, who did this sort of work, I get the feeling Maz is pulling rank on me. He was a *real* miner, not someone who sifted through river gravels to see what had tumbled down from the mountains.

'Coal truck,' Maz calls out as a truck goes past. 'Probably delivered a load to Canterbury – to Fonterra, hospitals, that sort of thing.' I guess it's some of the 'shit coal' – not good enough for coking, but decent at providing heat.

Maz drives us to a bakery that has been here since 1874. It's packed. There's no vegetarian option on offer, so I go for a chicken and mushroom pie and Maz orders the steak and cheese.

Next stop is the Reefton School of Mines, an old wooden building with creamy yellow walls and red roof and doors. Maz's great-great-grandfather, Henry Lawn, came here as a gold miner, then her great-grandfather Charles, Henry's son, went to Reefton School of Mines and became a miner and a mine manager. At some point Charles left and went to Te Aroha, then he, like Maz, transferred from mining to civil engineering. 'Same journey as me, I'm his reincarnation,' she says. Her ancestors were smart, she reckons. 'They got out of the cold smoky coal hole of Reefton and went north, where it's warmer.'

Reefton's Blacks Point Museum, founded by Henry Lawn and three other local men, is in an old Methodist Church. The museum is dedicated to prospecting and mining, displaying rocks, rusted machinery, books. There's domestic stuff too, ancient cans of strawberry conserve, OXO cubes, flavoured orange cordial and colourful rusted tins of 'almond cakes' and 'Pontefract cakes', all displayed neatly on white shelves. There's even a row of glass bottles, filled with gravels and mud layers brought up by the 1929 Murchison quake. The walls are lined with posters and news clippings of significant mining or geological events.

I notice there's a lot of 'pioneer' stuff in Westland. Maz and I talk about how we're descended from these people they call pioneers – nineteenth century miners, barmaids, governesses. They were Methodists, Anglicans, Rationalists and Free Thinkers. They were hard workers and progressive voters. But the word 'pioneer' suggests they were among the first settlers to arrive on the Coast. A more accurate word for what they were is colonists. In seeking better lives for themselves they were part of a system that did generations of harm to Māori, and to the land.

—

We drive into Greymouth in the afternoon, the flags on the road into town flapping in the wind.

'Look at that!' says Maz. 'The Barber is hooning.'

We're looking for a motel, but I'm also looking around, curious about the history of the town, the buildings that have been here since the nineteenth century. Marian and Edward were married in the Holy Trinity Church in 1871. It's now an 'evangelical-charismatic' church, with a Friday-night youth group, a monthly 'ladies fellowship group', and livestreams of the Sunday services on YouTube. I feel no need to visit.

We park up so I can google accommodation. A big sign on a wall by a grassy lot has giant letters *COAL GOLD TIMBER*.

I go for a closer look. *Our COAL our GOLD our TIMBER fired the furnaces that industrialised New Zealand, financed this country's growth, helped build this nation.* The logos in the corner are for Heart of the West Coast and the Māwhera Grey District Council. They're certainly owning it, I think. Even though tourism – pre-COVID, at least – is a bigger money earner than the extractive industries now, this is the story that persists. After checking various options – I'm not staying in any brick earthquake hazard building – we book into a bland-looking motel, where most of the other guests are uniformed police, then head into the library.

The stories passed down my family are all stories of European settlers, white people, except for my grandmother's best friend, Mary, who is in a photo I have and in stories my mother told me. In my grandparents' house there was a small kahu kiwi, a simple kiwi feather cloak, the feathers woven into a hessian flour sack backing, a gift from Mary's family to my grandparents, perhaps a wedding present. It hung on the wall in the spacious hallway, a grand entrance to an otherwise modest 1920s Christchurch bungalow. Across the hall from the cloak was an impressive set of mounted antlers, a 12-pointer that my grandfather shot, the skull of the red stag dressed in dark green velvet, and – inexplicably, and I regret that I cannot now ask my grandparents about this – beside that, an imposing portrait of eighteenth-century Māori leader Te Rauparaha.

I know that Māori Creek was the home of my great-great-grandparents, Edward and Marian, my own Nanna's Grandma and Gympie. But from what I'm reading in the library, Māori Creek looks like a Chinese mining community. I want to find out more. There's a folder labelled 'Greymouth cemetery' with lists of names and dates. I text my mother. Anglican? Yes. And then I find them. Garland, Edward, died 1932, and Marian died 1921. They're in the Old Anglican Block, row 17, plot 11.

I look up the Garlands in a book with the handwritten title 'Pioneer Family Register for the West Coast'. The register contains details of Edward and Marian's lives that I recognise from a family history written by my great-aunt, my Nanna's older sister. 'Sent to NZ by father's second wife. Death falsely advertised in Hobart newspaper. Daughter of a Governor of Tasmania.' These are the stories I've grown up with. How her stepmother put young Marian Milne, just a teenager, on a sailing ship to New Zealand from Tasmania, after tricking her into thinking she was on a ship home to England. But I've googled Tasmanian governors and never found one called Milne. It was a bit of a relief: I have enough white settler discomfort without finding an ancestor involved in the colonial Tasmanian government.

⁓

I wake early. Once it starts to get light, I walk down the road, the Barber blowing coldly at my back. Past the New World supermarket where trucks are unloading produce. Past quiet houses, along streets with European names like Tasman, Nelson, Chesterfield. Past a jogger, an old man walking with a stick, a mini-golf sign. The kind of neat orderly houses and properties that I always found so oppressive about Christchurch, the only place I've ever lived on the flat.

When I turn towards the west I see something blue at the end of the road. Without my glasses I can't see if it's a blue roof or the sea. I walk towards the blue until I can hear the roar of the waves, and over a sandy rise, to a wide grey beach. The rocky beach is empty except for a tepee built of driftwood.

Young harakeke are planted along the rise behind. I walk back to the cemetery, past a tūī in a yellow bottlebrush tree, past a small grassy airstrip and a TOP 10 Holiday Park, where there are no cars in the 'guest parking' spots.

The cemetery is on a grassy slope facing the mountains, and I enter through an old wrought iron gate. The graves are old. Some of the headstones on a lean, some disappearing into the grass, most low to the ground but with a few crosses and pillars rising above them. I don't know which end to count from, so I count 17 rows first from the bottom of the slope then from the top. I can't find it. There are some graves missing, the rows uneven, and it's hard to tell what counts as a row.

I walk around, up and down the rows, until I see the name *Marian* then *Garland*. It's a simple grave. A long concrete slab at the top of which sits a raised, angled headstone. There are two concrete flower holders below the headstone.

'Hello!' I say out loud. 'You are my family.'

Marian was 74 when she died in 1921, a good age for a woman born in 1848, and Edward senior, my great-great-grandfather, died in 1932, when he was 88 years old. But the headstone is not just for Marian and Edward. It also carries the names of five of their nine children.

Marian sailed from Hobart to Lyttelton in July 1864, aged 16, on the barque *Chrishna*. She was 'looked after' by a family, says my great-aunt's story. The *Lyttelton Times* reports the arrival of a 'Mr. and Mrs. Smith, five children and servant, in the cabin' in a sailing of the *Chrishna* that arrived in July. Given the lack of any other single women on the small passenger list, I think this – the 'servant' – is her. At some time in the next seven years, she found her way over the mountains, or around the coast, to the West Coast where she worked as a governess, and in hotels making beds and tending bars, and perhaps that's where she met my great-great-grandfather, back in town to celebrate with a drink after bringing his gold in to sell. She married Edward Garland in 1871, and then moved to the town of Māori Creek, where they settled, carrying all their belongings by packhorse along a rough track from Greymouth. She bore nine children and three of them died in childhood – one from pleurisy and convulsions after a six-week illness, one from meningitis, and one from intussusception, a bowel obstruction, after doctors at Greymouth Hospital were unable to save her. Today, it's likely they would all have survived.

As well as Annie, who died aged six, Annie Florence, who died aged 16 months, and Bertha Elizabeth, who died at just six months old, the headstone acknowledges Allan Gordon, killed at Gallipoli in 1915, and Edward, killed in a mining accident in Australia in 1911.

I drop to my knees and cry. I sob for how hard her life was. Marian Garland. When I trace back my mitochondrial DNA, that's where it leads: through my mother Ruth, to my grandmother Gertrude, to my great-grandmother Lydia, to Marian. And before Marian, to her mother, Rachel Saunders, who was born in England, and sailed from London to Hobart with her family in 1835. Before that, I don't know.

I know from my own research now that some of the story my great-aunt wrote was not quite true. Maybe it was the story passed down by Marian, or concocted from some delusion of grandeur. The true story is worse. Marian's father wasn't 'Governor of Tasmania'; he was Archibald Milne, a sailor born in Banff, Scotland, who worked his way up from crew boy and seaman on Southern Ocean whaling ships to become captain of the barque *Cacique*, sailing between Hobart, Lyttelton and Melbourne. We knew that Marian's mother had died, but I now know her father died first. 'Suicide by drowning', read the headline in the *Hobarton Guardian* of 2 October 1852. 'Archibald Milne of the *Cacique* has put an end to his existence by throwing himself into the sea whilst in a state of temporary insanity.' Marian was four years old. Her mother Rachel remarried in 1860 but died in 1861, a few days after her newborn son. Marian's new stepfather remarried one year later. So the story that her stepmother put her on the ship to New Zealand could well be true, but by this time Marian had no living parents and no siblings. I wonder if she had aunts – the shipping records suggest that Rachel had sisters – and grandparents, but perhaps they had returned to England, or moved to other parts of Australia, or even to New Zealand. I don't know.

Jonathan and I sometimes talk about how our lives feel really hard lately, but they're nothing compared with Marian's. Though, which would be better? To have a hard life but know your children will have a happier, more stable, more prosperous life than you have? Or to have a good life, with physical comforts, material wealth, state-provided healthcare, a fulfilling career, but to live it expecting that things will get worse for your children, and your children's children, and to know that your generation had the power to change things but didn't?

I wish I'd brought flowers to the cemetery. I look around for something I can pick and put in the flower holders, and settle on two yellow poker flowers. Their petals are bright against the grey concrete.

Marian, as well as birthing nine of her own children, was – according to family lore – a 'midwife and a healer'. My great-aunt's story says she was a 'natural gardener' with much 'traditional wisdom'. She baked and crushed eggshells from her hens to add to soups and stews. She told her daughters to eat an orange before a dance to make their eyes sparkle. She gave her children Lane's Emulsion and Irish Moss. She stitched wounds closed with a sterilised sewing needle.

She didn't have a mother, or an aunt, or a grandmother, to teach her, and I wonder if she instead learnt from books, from her copy of *Enquire Within*, published 1873, with 'everything from choosing fish to a nosebleed, from mixing medicinal herbs to tapestry'. Or were there local women, Māori, European or Chinese, who taught her? My great-aunt's story says that Marian, midwife to the local community, delivered 72 babies, 'never losing one'. She also kept house, cooking on a colonial oven and in pots hanging from hooks over an open fire, 'keeping the beds spotless, the frilled covers freshly laundered daily' and sewing for everybody, 'even turning a coat for one of the Chinamen'.

Because of the remote mining town they lived in, I'd always imagined her and Edward in a very simple house – how could you build a house without a builder? – but my great-aunt's story says that Edward built them a fine house, pit-sawing all the timber, making beds and tables and cupboards, and fencing the garden with wooden slats. In the garden they grew vegetables and planted poplars, a cypress, apples, gooseberries, black and red currants, and a cherry tree. Cottage roses and nasturtiums climbed over the house, there were geraniums in pots, and daisies, pansies, forget-me-nots and pinks in the garden. After the children married and dispersed around the country, Edward and Marian lived there, with her seven cats, until they packed up and got in a taxi from Māori Creek to Lydia and Anders's place in Kaniere, with Desdemona – 'queen of all the cats' – sitting on Marian's knee.

'When Grandma came to live with us in her later years,'

tells my great-aunt, 'Desdemona would come to call Lydia when Grandma was in need, and the night Grandma died she came to Lydia pulling urgently at the bedclothes. After the funeral she disappeared and was never seen again.'

After my early morning walk to the cemetery, I return to the hotel, where Maz is up and dressed.

'You ready? Let's rock'n'roll,' she says. She's buzzing because she's just heard that her brother, who lives in Brazil, has finally had the COVID-19 vaccine. Things have been tough in Brazil, with few regulations and the virus surging, and she's relieved to know he and his family have been vaccinated.

Our first stop is the supermarket. After our week-long roadie, we're heading inland, for a quiet weekend at the Blackball Readers and Writers Festival. Laden with supplies, we drive out of Greymouth in the morning light, our first mission to search for the place where Marian and Edward raised their family. Edward came to the West Coast in 1864, from his home in Australia, attracted by the gold. He was a gold miner all his life, so he must have done okay. In her family history, my great-aunt remembers him, in his sixties, 'sitting at the end of the kitchen table with his delicate scales, weighing out his gold, gently blowing away any specks of dust, before taking it off to Greymouth to the bank'. From where he lived, Maz infers he was an alluvial gold miner, mining gold from river gravels.

Māori Creek, I've learned from the Greymouth Library, was renamed Dunganville in 1880 – though locals never called it that – and that's the placename that now shows on Google Maps. Dunganville is a 25-kilometre drive from Greymouth, and it soon feels remote. The road is windy and hilly, and we pass bush, farmland, forests. After 30 minutes or so I'm excited to find we're on Māori Creek Road. My ancestors feel close. We pass an ! sign in an orange diamond. Beneath it is a *MINING OPERATIONS* sign wonkily attached to a pole. *MINING* has been written over a previous word, *LOGGING*.

With the ghosts of family history swirling around me, Maz and I drive further inland, into a river valley, past wild blackberries growing beside the road, past a stand of eucalyptus trees, through a gorge along a road that is alternatively paved, gravel, paved. Along the roadside are some big trees – tōtara, rimu, pine – as well as scrubby bush and grass. We cross a one lane bridge, and the road takes us down a slope into a hollow, past a sign saying *8 Mile Creek*, and I realise this is where the original Māori Creek settlement used to be.

On 'the Clifton Terrace', the original site of the town, we find a bit of flat land with a north-facing aspect. It's a nicer location than Reefton, which is surrounded by hills, but there's no sign of human activity except a sign, white paint on plywood, *PRIVITE KEEP OUT.* In the late nineteenth century the town was moved upriver, and the old settlement 'sluiced off the face of the earth', according to historical notes I read at the library. We try to figure out exactly where the town was, but there's not much to see.

There's a gravel road near where we've parked and I walk up a track lined with gorse, ferns, mānuka and exotic weeds and grasses to a wire mesh gate on a fenced-off bit of land. A sign on the fence says *MINING SITE: UNAUTHORISED PERSONS KEEP OUT. THIS IS A MULTIPLE HAZARDS AREA.* Over the fence is a gravelly clearing. I can see a few big black truck tires, marking off an area, a yellow digger next to a mound of rubble, a three-sided wooden shelter with a corrugated iron roof that would have looked in place 100 years ago. There's a puff of smoke in the distance.

'Is this what gold mining looks like today?' I ask Maz when I'm back at the car.

'Looks like they've cleared the ground to do a bit of alluvial mining,' she says.

The ground is lumpy and uneven. All the land here is disturbed, 'probably for the second or third time', says Maz. I realise it's futile trying to read the landscape, use my knowledge of geomorphology to interpret where the river was, how this landscape formed.

'So as the technology changes they go over it again?' I ask.

'Yeah, or if the gold price goes up it becomes worth it again.'

We drive further up Māori Creek Road until we lose signal, but I keep navigating with Google Maps, trying to get to the spot marked Dunganville.

We drive on through farmland – we see black and white cows, pairs of Paradise ducks – until we realise we've gone past, or through, Dunganville without even noticing. There's nothing there. We pull off into a gravelly road through pine trees, in the hope of a space we can turn around, or because it looks like a loop track back to the main road. It's pretty desolate and unkempt.

'Now's about the time we're going to get abducted and murdered,' says Maz.

'Sssh.' I don't want to think about it. 'The map says there's a right turn here . . .'

We turn back and drive slowly down the gravel road. We see a dwelling of sorts. 'Probably a marijuana-growing operation or a mining shed,' says Maz. Further along the road is a sign saying *MINING OPERATION*. Closed gates. *Ngāi Tahu forest estate, entry by permit only*. A portaloo. The hills are denuded and newly gorsed.

We re-meet the paved road and head back the way we came.

Cows are watching us. There's some metal stuff beside the road, hard to know if it's junk or archaeology. Some railway sleepers.

At the spot that I guess is the closest to Dunganville we're going to get is one cute little house surrounded by farmland.

'This place is spooky,' I say. There are too many locked gates, empty lots, disturbed histories. It feels abandoned, neglected, and I don't much like it. As we head back to Greymouth, we talk about what it must have been like to live here 150 years ago, before there were roads, before there was electricity, running water, sanitation.

'They must have been rugged as, those women,' says Maz. 'Fuck yeah.'

⁓

We're done with our exploring and are coming to the end of our trip. Blackball is one of the West Coast placenames I know from my childhood. I remember my mother talking about it, telling stories of my grandfather visiting as a school kid, when his best friend Ralph Glasson, who boarded at Christchurch Boys' High School, returned there for the holidays. His family made honey, and still do.

We follow the Grey River inland, along the north bank of the river. Ten or so minutes out of Greymouth we drive past the Brunner Memorial, which commemorates the 1896 underground mine explosion that killed 65 men.

'The worst industrial accident in New Zealand history. You think we would have learned something from it, but we didn't,' says Maz.

Intergenerational memory is tricky. It can be destructive, responsible for decades-long grudges, wars even. But it can also be helpful, for acknowledging your ancestors; for remembering pitfalls, dangers, and warnings; for better understanding the troubled society we live in today.

About 15 minutes into the drive, we leave the river and head north towards Blackball. If we carried on inland, then turned south rather than north, we could make our way to the Gloriavale Christian Community, where the women's high-necked, full-skirted dresses look a bit like our 1980s choir girl get-up. As we approach Blackball, I check my phone for instructions. We're staying in the Old Post Office, an Airbnb booking. It's easy to find, 'in the middle of town', says Gary, who has messaged us the code to the lockbox. I look it up as we drive into town.

'Do you require one queen bed or two?' I read out a message from the day before. 'Jesus,' says Maz.

'Two!' I reply, hoping he receives it in time. It's been fun sharing a room for the last few nights, but I think we'd draw the line at sharing a bed.

We drive into town along Blackball Road, the middle of a large area of flat land surrounded by high rugged hills, including the peaks of the Paparoa Range. The valley looks to be populated by quarter-acre sections with small wooden houses, coal smoke hooting out the chimneys.

On the outskirts of town, a long high decrepit fence is covered in colourful hand-painted signs. We slow down so I can read them. *1080 Poison Kills KEA FISH INSECTS BEES AND HUMANS AND IT'S IN OUR DRINKING WATER! POISONING OUR CHILDREN* says one. *YOU ARE WTINESSING AN EXTINCTION EVENT. DEATH, TORTURE, CRUELTY BY 1080 ENFORCED BY NZ POLICE + GOVT* says another sign, with pictures of dead kiwi, carked out, upside down. *WELCOME TO OUR POISONED*

NATION CHEMICAL WARFARE ON ALL CREATION is on a sign featuring a red skull-and-crossbones and a dead tree, fish, and kiwi. Down the driveway is a white 4WD with slogans painted on the sides. *STAND UP . . . FIGHT BACK OR DIE A POISONED SLAVE* and *STOP THE DROPS, LOCK UP DOC.* I'm intrigued and would quite like a chat with whoever made these signs. What's driving these passionate feels about 1080? But we have a pōwhiri to attend.

We find our Old Post Office accommodation near the centre of town, and Maz points out other notable buildings – the mine manager's residence, the jailhouse, the pub across the road, a pie shop.

We park, unload, light the fire in our digs to warm it up, then walk up the road in the rain to the school, where the festival will be held. We join a throng of people, mostly older than us, gathering under the shelter of a porch, waiting to be welcomed into a classroom. People are greeting one another and gathering around a table laden with tea and coffee and homemade cake. Maz has been to the festival before, knows a few people – mostly of our mothers' generation – and pours herself a cuppa and starts chatting until it's time to go inside.

'Blackball's getting ahead,' says Paul, the organiser of the festival, when we bump into him after the pōwhiri, and Maz asks how he's been since she was last here. House prices are ridiculous now, he says. 'What would have cost $90,000 a couple of years ago is now selling for $190,000. The Aucklanders are buying up Greymouth.'

We talk about the impact of COVID-19 on the regions. Paul doesn't like what our COVID response has done to New Zealand. It's saved lives, which is great, but it's also made us too insular, too self-congratulatory, he says. 'It's like the 1950s.'

We're enthused about the book festival but after a week of being on the road we're not sure if we're ready for meaningful social interaction, so after the pōwhiri we retreat to (Formerly) The Blackball Hilton, a two-storey wooden hotel, built in 1910, offering food, drinks and accommodation. We arrive at the bar just as a man in black jeans and T-shirt is finishing a bagpipes solo, after which a group of middle-aged men – we think they're on a motorcycle tour – clap and cheer.

In our trapper hats, coats and check shirts, we don't really look like our Auckland and Wellington selves, but it feels right for this trip, for who we are right now. We unload our coats and hats at a booth table, order a couple of beers from the bar, and play pool while drinking from the bottle. One of the blokes in the pub looks like he's coming over to talk to us, but I see his mate shake his head, *no*.

When we're done playing pool, we explore the walls of the pub. Around the corner from the bar and pool table, behind the eating tables, are wall displays about the town's coal mining history, the 1908 Blackball Miners' Strike, the 2010 Pike River disaster, and the recent opening of the new Paparoa Track. There are framed signs, I assume from the strike – *No Scab Labour*, and *United We Stand, Divided We Fall* – and a yellow and red hammer-and-sickle T-shirt for sale. In the hall there's an oil painting of David Lange, and we take a photo of ourselves smiling underneath.

Afterwards, we join the festival crowd at the Workingmen's Club, where I order a pint of local beer and get chatting to a couple of farmers about our road trip. There's a dog nearby, demanding pats. A woman in a chair knitting a cardigan. A dartboard. On the wall, there's a photo of the gentle smiling face of Michael Joseph Savage, the same photo Maz had on her living room wall when we were teenagers. The same photo

Jacinda Ardern had on the shelf behind her when she made her March 2020 announcement about New Zealand going into lockdown in response to the COVID-19 pandemic. A photo of the Prime Minister who – I learned in fifth-form history classes – introduced social welfare to New Zealand, made this a better place to live.

I think about how joining a Pentecostal church was about the most rebellious thing we could have done as teenagers. It was the one thing my mother explicitly directed me *not* to do, warning my purple-haired, pot-smoking teenage self about the Pentecostals who sat around speaking in tongues and other such nonsense. I wonder how much our urge to rebel against our parents affected our choices? But for Maz and me, it became so much more than rebellion.

Looking back, though, in the light of today's evangelical churches and the criticisms levelled at them, I didn't see anyone getting rich from our church. But I did see a culture of conformity and the perpetuation of the patriarchy. The world was run by men and there was no pretending otherwise. Jesus was the head of the Church, men were the head of the family. And it was men who led the church; there were no female pastors or evangelists. We dissed the big Wellington churches for being fake and impersonal, too American. While us teenagers got caught up in End Times fantasies, and our pastor started leading us on a bigoted path that was anti-gay and anti-women, most of the people at our church were humble, kind and true. Hutt Valley style.

—

'You again,' says Maz the next morning as she walks out of her room in pyjamas. I've already had a shower – I put brown conditioner I bought from Lena in my hair, might start making an effort again, I think – lit the fire, made tea, and opened the curtains so we can look at the rain.

After breakfast we walk up to the school in time to see Becky Manawatu discussing narrative with Stevan Eldred-Grigg, who seems to be talking quite a lot about his own writing and the people he's met. While I would later seek out one of Eldred-Griggs's books, *Diggers, Hatters & Whores*, to find out more about the local history, today I just want to hear from Becky, who's talking about her novel *Auē* and her own upbringing on the West Coast. I'm intrigued when she starts talking about being attracted to Girls' Brigade, Sunday School, then being baptised. She was drawn to the stories, she speculates.

My sister joined the Girls' Brigade for a while too. And I remember the large hardback *Illustrated Bible for Children* my Methodist grandmother gave to us in an attempt to give us some sort of Christian education. I loved it, and returned to it again and again for the stories – the parting of the Red Sea, Moses and the Burning Bush, the Loaves and the Fishes.

Next up is Nicky Hager, who is in conversation with Paul Maunder about his books *Hit & Run, Dirty Politics, The Hollow Men* and more. I'm knitting while I listen – it helps me to focus – but I soon put down my knitting and begin to write down his words. 'Growing up feeling different is a curse when you're a child,' he says, 'but a benefit when you're an adult.' This resonates with me. Maz tweets that he's her new crush.

I first noticed Nicky Hager in the 1990s, when he was a spokesperson for the anti-nuclear peace movement that so many people I knew were involved in. 'Real politics is trying to change the world through democratic action,' he says now, and I write it down.

The next thing I write in my notebook is, 'The difference

between persuasion and propaganda is deception.' Were we deceived, I wonder? Did the evangelists and pastors and elders really buy into the stuff they were teaching us? Or did they have ulterior motives? Was someone or something behind all those End Times narratives that had us simultaneously euphoric and terrified?

'We're here to work as hard as we can, for as long as we can, on the things we care about,' Nicky is saying.

This resonates with me. Whenever I find myself lost or unhinged, throwing myself into my work – whether it's a book, a gardening project, a teaching plan – is how I calm myself. But 'the things we care about' means different things to different people. How can we commit ourselves to the things we care about without making it weird? Without getting so caught up in our own issues that we demand everyone share our views? I think about an essay I read by American writer Rebekah Matthews. 'I don't like the idea of people united by their shared certainty,' she says. 'I like that I am able to look at the world with nuance, an interest in complexity, an understanding that there is no universal roadmap every person needs to follow. When I feel lost sometimes, I wonder if that's what it means to be human.'

I like what she said, but there are some things we need a bit of shared certainty on. As we face a future increasingly impacted by climate change – with associated sea-level rise, wildfires, and floods – there are things we need to agree on if we want our civilisation to continue.

⁓

We end our trip and drive east, over the mountains to Christchurch, where I will drop Maz at the airport and join my daughter for a continuation of my roadie, to Queenstown this time. As we drive, I think about family, my son and two daughters, and my maternal ancestors. And I think about how, on the last night in Blackball, Maz and I sat on the back porch for a while, enjoying the rain. How, back inside with a bedtime cuppa, Maz, still in hat and coat, pulled her chair close to the wood burner and played her harmonica while I sat on the couch across from her, thinking and writing.

At the start of the pandemic I'd used the lockdowns as an opportunity to withdraw, shed what had become my costume, my many masks, and had started to re-engage with what I thought of as my feral self, a more essential and honest version of me. Now, after a week on the road, talking, listening, laughing, I can see that I needed to find a way back into the world, try once again to figure out my place in it. A place where I could be honest and useful but not consumed by anxiety. A place where there was more room for laughter. I needed to unleash a stronger, calmer and more vibrant version of myself who I had a feeling was there, underneath all of my anxiety about the state of the world. Writing seemed to be a way to find her.

Notes

This book is a work of nonfiction. For the 2021 strand of the book, the people who I interviewed and whose words and names are in this book have given me permission to quote them.

Occasionally, I have used words from a person I met but whose name I didn't catch, or whom I was unable to contact for permission. In these instances, I have made up a name. In the teenage portion of the book, I have protected people's individual identities by using a range of generic names (Paul, Mark, Simon, Amanda, Billy, etc.) or by grouping experiences (our mothers, the Christians, etc.).

A note about my Christian experience: This book records a very personal narrative of my involvement in an evangelical church in the 1980s. In sharing my story I don't claim to speak for anyone other than myself (and, to a lesser extent, Maz).

This book includes quotes from unpublished family histories written by my great-aunt, Marion Dorrington (née Hackell).

The quotes on pages 13 and 194 are from the GNS Science Consultancy Report 2016/33, 'A Natural Hazard Assessment for the Township of Franz Josef, Westland District', by R. M. Langridge, J. D. Howarth, R Buxton, and W. F. Ries, pages 33 and 37 respectively.

On page 25–26, the description of the Murchison earthquake quoted from Te Ara Encyclopedia of New Zealand is originally from 'The Murchison Earthquake, 1929', an entry in *An Encyclopaedia of New Zealand* edited by A. H. McLintock and published in 1966.

The research mentioned on page 194 featured in a RNZ news report published 20 April 2021: 'Alpine Fault: Probability of damaging quake higher than previously thought'.

The quote on page 195 is from a second RNZ report published 20 April 2021: 'Franz Josef emergency services on Alpine Fault must be moved – Westland mayor', by Eleisha Foon.

On page 214, the quote about Archibald Milne comes from a news item on page three of *The Hobarton Guardian*, 2 October 1852.

The Rebekah Matthews quote on page 225 is from her essay 'A softer answer' in *Empty the Pews*, edited by Chrissy Stroop and Lauren O'Neal.

Acknowledgements

First, I would like to thank my husband Jonathan King, for being an excellent critical reader of my early drafts, and for solo parenting while I was away on writing retreats and outdoor adventures. And big love to my wonderful children, Pippi, Hazel, and Huck, and thanks for your enthusiasm for the book cover.

I started writing this story in February 2021, when I took my teenage journals to the Kāpiti Writers' Retreat as inspiration for something I was thinking of writing about my two years as a born-again Christian. In her workshop 'There's no "I" in Essay', Rose Lu gave me the prompt I needed: the idea of writing in the first-person plural was what got me started on the teenage narrative that is one thread of this book. Thanks to Rose Lu, and my fellow workshoppers who, after I read my 743-word essay aloud the next day, encouraged me to keep going with this idea. Big thanks also to Kirsten Le Harivel and the rest of the team at Writers Practice Aotearoa.

Later that year, I enrolled in a six-week online Catapult workshop led by Frances Badalamenti: 'The Study and Practice of Autofiction.' While I was writing *nonfiction*, I wrote a first draft of the teenage narrative as weekly submissions for this course. Huge thanks to Frances for introducing me to so many wonderful autofiction writers, and for – along with my classmates – providing feedback on my work.

While one thread of this book focuses on my teenage years, the other thread covers a 2021 road trip down the West Coast where I enjoyed meeting and listening to locals. First thanks to the incredible generosity of Tom and Emily in Westport, and Milly and Steph in Karamea, for having us to stay and for talking to me so openly and entertainingly. Thanks also to Bruce Smith for letting me interview him and include his words in my book. Also thanks to Aidan and colleagues at the Seaview Lodge in Hokitika. And to everyone else along the way.

After spending most of my writing life without a writing group I'm now lucky enough to be a member of two: big thanks to the Glen Road Writers and my US online writing group for feedback on various parts of this narrative. Special thanks to Elizabeth Knox for feedback on a full draft of the interwoven story.

At Te Herenga Waka University Press, huge thanks to Ashleigh Young, Anna Knox and Todd Atticus, and to the rest of the team involved with this book: Fergus Barrowman, Craig Gamble, Tayi Tibble and Ruby Leonard.

Heartfelt thanks go to Marianne Rogers (Maz), for helping me navigate a world full of humans and other hazards, and for letting me tell this story on behalf of her too. Thanks also to our mothers, Margaret Rogers and Ruth Brassington. I wish I could go back in time and tell you that we would both turn out okay.

Finally, I acknowledge my teenage self, for keeping such detailed diaries, and my later self, for holding on to these diaries despite the cringe factor. Sometimes hoarding pays off.